My Grandfather, Clayton Cawley, with a Pringle Nitroglycerine shooting truck near Eldred, in the Bradford Field. The derrick and drilling were by the Cawley Brothers, who once drilled and operated many wells in the area.

The Bradford Oil Basin

A Regional History of Oil Technology

Dr. Jon C. Cawley

Sometimes Fellow of
The Smithsonian Institution

1986, 2019   Dr. Jon C. Cawley

The primary content of this book was first produced and published as an Undergraduate Senior Thesis at the Pennsylvania State University, in 1986. Its substantive contents are essentially complete here from that edition. The original Undergraduate Thesis copy remains on file in bound paper form at Pattee Library on campus at State College, Pennsylvania. This edition adds Cawley family geneological information that is new and original to this volume and edition. Historic photos and post cards are primarily from the collection of Caroline J. Crain.

All Rights Reserved.

Published in the United States of America
in 2019
by Ravens Table Press co-op

Printed in 11 point
"My Underwood" font

Library of Congress Cataloging-in-Publication data

Jon C. Cawley

The Bradford Oil Basin
A regional history of oil technology

2019

RESURGAM

"A small house agent's clerk, with one bold stare,
One of the low on whom assurance sits
As a silk hat on a Bradford millionaire."

-T.S. Eliot, The Wasteland, 1922

This book is respectfully dedicated to my great aunts, Rosamond and Coletta Cawley, both of whom spent more time with this manuscript, and were more appreciative of the living role of the Cawley family in the history of Bradford Oil than anyone else.

The Bradford Oil Basin

Early Standard drilling rig at Indian Creek, near to Eldred. Bradford Oil Basin.

Jon C. Cawley

Introduction

The Bradford Oil Basin was the first large oil-producing basin in the world. Discovered in 1871 (the year of the Great Chicago Fire), the Bradford region quickly achieved distinction as the high-grade oil capital of the world. The tremendous output of Bradford oil soon changed the structure and logic of the oil industry completely, and laid the foundations of our Oil Age of the 20th Century.

Earlier wells of the older, more southern Pennsylvania fields were already well established by the time Bradford fields began to produce. Prior to 1871, oil was little more than a cottage industry, albeit a successful one. Teamsters carried crude in leaky barrels to city markets where it was sold as a patent medicine or specialty lubricating or lighting oil. Production was unsure, and the markets were hardly stable; the business was one of "boom and bust". Oil towns appeared and disappeared almost overnight, as the cycles of scarcity and oversupply had their inevitable result.

Bradford basin oil, however, soon showed itself to be prodigious and apparently inexhaustible. By the 1880s, Bradford production and overproduction had changed petroleum products from specialty items, to the ever-present and ever-necessary commodities that they are today. In the subsequent years, Bradford basin fields have retained their unique position in oil history, by means of their persistent growth and development in the production of Pennsylvania Grade Crude Oil.

The Bradford Oil Basin

From the Bradford region has come an astonishing amount of oil technology and application. Bradford basin producers, from the start, have been leaders in oil technology. This technology has encompassed changes in derrick design and usage, well drilling, and well management. Important has been Bradford's contribution to the science of oil storage and transport from the fields to distant markets. Significant too, has been the saving of the Bradford fields by pumping and water flooding; paving the way for research into tertiary recovery. This study presents an in-depth look at these technologies and their use in the Bradford region.

No history, however, is complete if it deals with technology only. The story of the Bradford oil basin is a story of the North Central Pennsylvania wilderness as well. It is a story of a hundred small oil towns built among the derricks, a story of oil drillers and oil pioneers. It is the story of Bradford, the "Hub of Oildom." In addition, it is the story of the rock record, of geology and ancient seas: the origin of Bradford's strange, green paraffin-based petroleum.

Pavilion and Park at Rock City, near Bradford, Pa.

Jon C. Cawley

42. OBSERVATION POINT AND ENTRANCE TO ROCKS, ROCK CITY PARK, BETWEEN OLEAN, N. Y. AND BRADFORD, PA.

Chapter 1 Geography of the Bradford Field

The Bradford Basin

"The Bradford Oil Basin" is an area of major oil production in north western Pennsylvania and southernmost New York State. In its broadest sense, the "basin" is the upper extreme of the northeast- trending Appalachian oil belt that extends from West Virginia and Ohio through Pennsylvania.

The region includes the fields around Bradford City, that draw oil from the productive "third sand" reservoir below. This Bradford oil pool is situated within the northcentral area of McKean County, Pennsylvania. It extends north a short distance, into Cattaraugus County, New York, with about fourteen percent

Another Standard drilling rig at Indian Creek, near to Eldred. Bradford Oil Basin.

The Bradford Oil Basin

of the field being in New York State. The entire Bradford reservoir occupies approximately 84,000 acres. Job Moses first successfully extracted oil from this major reservoir at Allegany, New York, in 1871. Drilling proceeded southward, and wells of significant production volume were struck around Bradford by 1875.

Beyond the edges of the Bradford reservoir, the "third sand" is less productive. In some areas, the sand is flooded with salt water; in others, it is hard and dense, containing neither oil nor water. However, to the south and east of the main Bradford reservoir, there are smaller detached reservoirs of oil or gas, which lie within the sand.

Thus, regions to the south in McKean County and an area to the east in Potter County are also part of the producing basin.

To the northeast of Bradford in Cattaraugus County, New York lies the smaller Richburg pool at Bolivar and Richburg. Opened during a small "boom" of its own in 1882, the Richburg field is also stratigraphically equivalent to the Bradford third, and can be included in the general region.

To the west of Bradford lie the various small reservoirs of the Warren County fields. These extend in places into Allegany County, New York, to the north. These reservoirs to the west seldom lie within the third sand. Instead, most of these fields represent productive lenses of smaller, less important sands. Because this area is geographically related to Bradford, and because the non-third sand production is relatively minor, these fields can also be considered as part of the general Bradford basin.

The oil basin is centered at Bradford itself. The basin consists of a myriad of small oil towns and developments scattered across the forested hills and valleys.

Jon C. Cawley

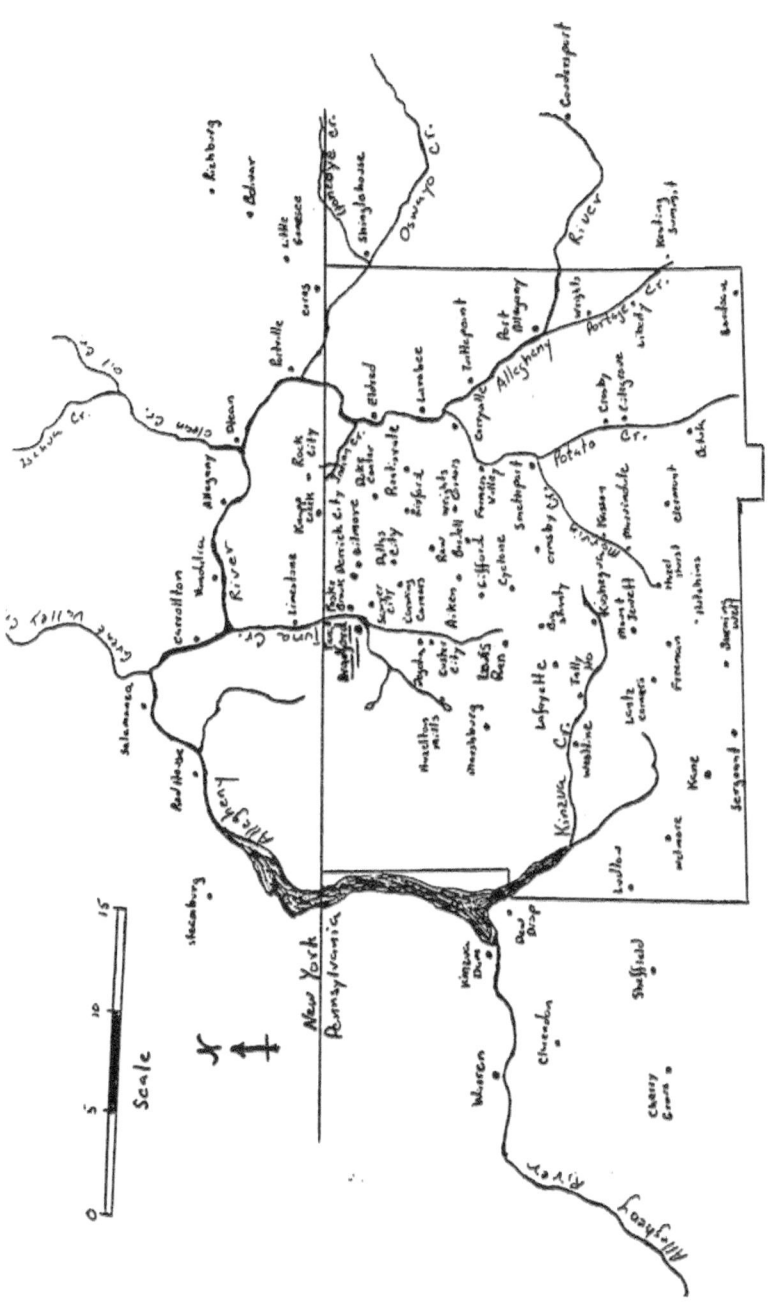

The Bradford Oil Basin

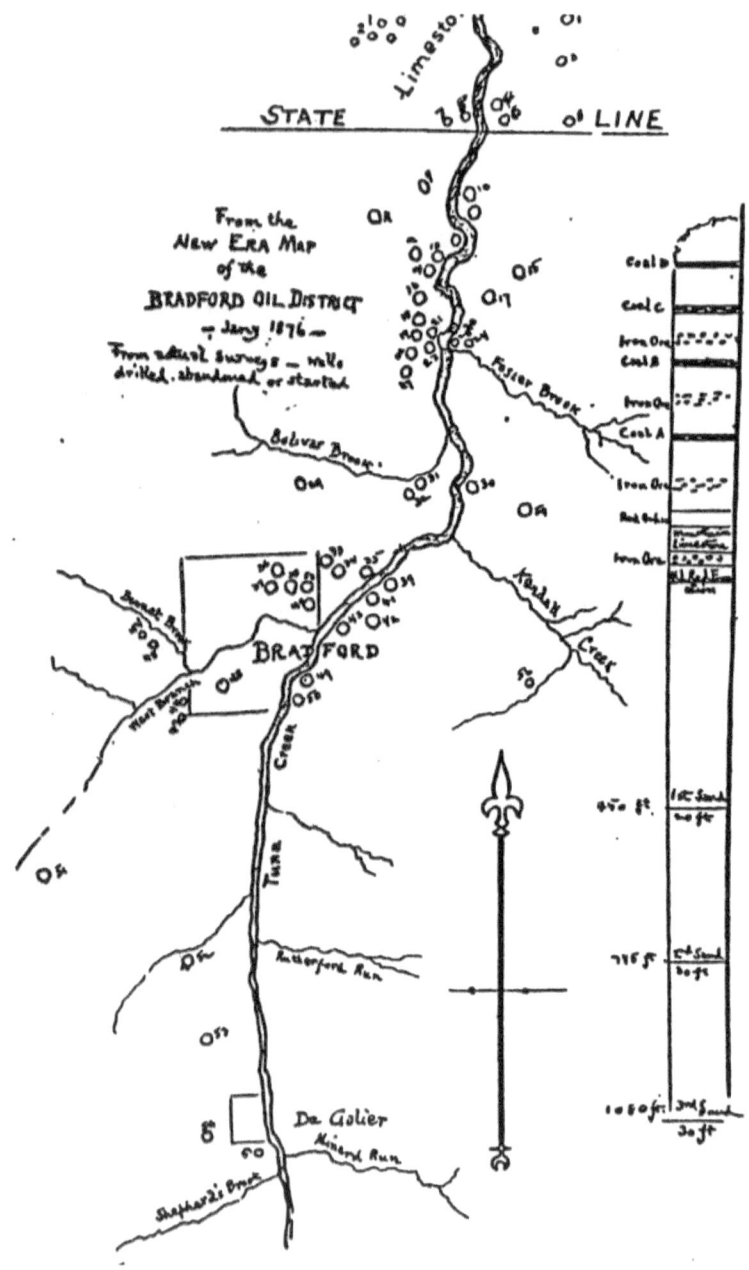

The region is fairly isolated; Bradford is about seventy-eight miles from Buffalo, about forty miles from Jamestown, New York, about one hundred thirty miles from State College, Pennsylvania, and one hundred sixty-three miles from Pittsburgh.

The northwestern area is a sparsely populated, thickly wooded high plateau region of rolling hills and winding, streamcut valleys. The area, centered in McKean County, is a portion of the Allegheny Plateau Province. To the south around Kane, the upper highlands form the "Big Level," relatively flat, and intersected by deep, narrow valleys. The north edges of this highland are scattered with broken boulders of the underlying Knapp Creek sandstone.

The area here is best known for lumber; the stands of hemlock, pine, and hardwoods have long been considered as "The Black Forest of Pennsylvania." Kane is primarily a lumber town, with a history of sawmills and narrow gauge railroads.

To the north, the low, flat-topped hills of McKean County give way to the wide, gravel-filled river valleys of New York State. This area, north of Salamanca and Olean was overridden and worn by glacial ice. The wide, U-shaped valleys extend north to the Great Lakes and the vineyard and orchard lands of upstate New York. In the northern end of the field, the gentler topography is better for railroads and pipelines, and Olean was soon a prime shipping point for Bradford basin oil.

The Bradford Basin is drained by the Allegheny River system. From its headwaters east in Potter County, the river system runs northwest into New York State. There, its course changed by glacial blockage, the river makes a wide arcing turn and pushes south again to flow into the Ohio at Pittsburgh. The Allegheny marks the

The Bradford Oil Basin

general edges of the Bradford reservoir, surrounding it on three sides. In times past, it was important in transportation and shipping.

For Indians and early settlers, the river was a migration path into the Western interior. Port Allegany was formerly known as Canoe Place, and was the portage point between the Susquehanna River basin to the south and the upper waters of the Allegheny. Later, the lower waters were used for shipping. Lumber was floated to markets in Pittsburgh and beyond. In addition, small steamboats were known to come as far north on the river as Portville in seasons when the water was deep enough. Because of its glacially filled river course, the Allegheny is a prematurely old river in the Bradford region.

The "river of the Allegwi" wanders laterally in its river course, making curves and oxbow lakes. Between Eldred and Olean it has well-developed natural levees and back swamps.[1] During the years when the surrounding land was clear-cut, and erosion was increased, the river silted up considerably and the levees were accentuated. In some places today below Eldred, the level of the river channel is actually higher than the back-swamps on either side.[11]

Above the river at Indian Creek, the hills are topped with the boulder fields of Rock City. The huge stone blocks are jointed and broken Olean Conglomerate, which forms the capping bedrock. The hills at Rock City form the dividing range between Bradford and the river and shipping points at Olean and Salamanca. Early oil pipelines were built up and over the Rock City hills. When oil was discovered in the Indian Creek valley, the boulders of Rock City competed for space with the oil derricks.

Two early post card views of Rock City at the top of the Indian Creek Valley. c. 1906

The Bradford Oil Basin

Jon C. Cawley

Views at Indian Creek,
McKean County, Pennsylvania.

The Bradford Oil Basin

The Bradford Oil Basin

Indian Creek, a tributary of the Allegheny River, McKean County, Pennsylvania.

Jon C. Cawley

The Bradford region was originally covered with a dense forest of both coniferous and deciduous trees. The shale hillsides and valleys were filled with hemlock and white pine, beech, maple, cherry, and birch. These stands were often heavily undergrown with dense thickets of mountain laurel in the valley bottoms. The sandstone hilltops supported populations of chestnut, maple, and oak. Hickory and elm were also common on the hillsides and ridgetops.

The region's wildlife population was noteworthy, as well. The Allegheny River valley wilderness area supported large herds of white-tailed deer and elk. Bear, fox, bobcat, and eastern mountain lion were common. The northern valleys were known for beaver, and the Big Level area was home to the eastern 'lobo' wolf. Streams throughout the region contained large trout populations.

The Allegheny River was populated by bass, muskellunge, trout, and snapping turtles. Hawks and eagles claimed the sky above the region.

The area around McKean County was geographically impressive, but was not especially hospitable. It was used by the Iroquois as hunting ground, but there was almost no settlement. The region was dark and densely wooded, with ferocious animals, as well as game. Early European explorers had few high opinions of the "forboding wilderness." Steven Brule', who visited the Huron tribe of southern New York State in 1615, described the region as:

"Thick and almost impenetrable forests, woods and brush, marshy bogs, frightful and unfrequented places and wastes." (Brule', 1615)

In the western and southeastern portions of the region, rattlesnakes were a problem. The hunter Philip Tome, writing circa 1800, told of rattlesnakes along streamcourses being so numerous in places that one could not beach a canoe on the shore. With snakes, prowling

The Bradford Oil Basin

bear, panther, and howling wolves for company, the region would hardly have been pleasant, especially at night.

After the early 1800s, however, the forests were beginning to be cut for lumber. Available land was clear-cut; additionally, some was cleared for farming and settlement. Younger successions of plant life quickly filled in the cut regions. Goldenrod, blackberry, thistles, elderberry, chokecherry, and red maple were quickly common to the region. Wolves and panthers were eradicated, and elk herds were broken up. Niches were opened for new and different animal species as well. The cleared lands grew back as grassland and brush, offering cover for grouse, woodcock, and rabbits. The deer population thrived and grew in the new environment.

The cutting of the forests accentuated the cool, wet North Pennsylvania climate. The Big Level area, with its relatively high elevations, was especially affected by the clearcutting. Bare areas of ground tend to radiate heat quickly to the atmosphere, rather than to moderate temperature changes as a forested area would.

This "barrens" effect is quite severe around Bradford, making it the coldest area of the state. The town of Smethport in the region holds the official "record low temperature" for Pennsylvania. On January 1904, Smethport recorded a temperature low of minus 42. Temperatures of -10 to -20 Fahrenheit are not uncommon.

The daily temperature range is relatively wide as well, often exhibiting temperature swings of twenty degrees or more. Summer temperatures are occasionally above one hundred degrees. The growing season is comparatively short, usually between one hundred and ten, and one hundred and thirty days. The last severe frosts are early in June, and the first frosts of fall occur toward the end of September. It is a variable climate relatively unique to the area.

G 3895 View on Tuneangwon Creek, Bradford, Pa.

This is where I am visiting now. 1907

Tuneangwon Creek, Bradford, Pa.

View on the B. B. & K. Narrow Gauge R. R., Bradford, Pa.

Early post card views along the Bradford, Bordell and Kinzua Narrow Gauge Railroad.

View on the B. B. & K. Narrow Gauge R. R. Bradfort, Pa.

Jon C. Cawley

By the 1860s, McKean County was slowly becoming more civilized. Lumbering was progressing, and roads and railroads were being put through the valley. Population rose slowly; a pioneer economy prevailed in the region until the discovery of oil in 1871. Settlers in the region were hardy, and despite a major lack of conveniences and doctors, the population survived.

With the discovery of oil, the population increased sharply in a short period. Derricks quickly covered the hillsides, and oil flowed into the streams. Clay roads were driven into muddy trenches, and wooden plank roads were laid over top of them. Narrow gauge oil railroads quickly filled the region. There were "boom towns" in every valley of the Bradford basin. Still, despite the coming of civilization, the land still held much of its wild look. An article in the Bradford Era of a Bradford and Kendall narrow gauge train ride described the countryside in 1878:

"The train lays for a few minutes at the Kendall depot, then pulls out and rushes away, up the banks of the rushing, roaring Kendall Creek, swollen so much by the rain that at many points it washes the ties of the road on and up the broad, beautiful valley that even the destroying touch of oil operations has failed to make homely with its flashing stream, tail, stately hemlocks, its smooth meadows, smiling grain fields, happy homesteads, and over all, the peaceful quiet that reigns supreme. The trees and verdure freshened by the rain, stand out clear, bright and green, presenting all the varying shades of nature, while the tail, spire-like derricks, in the morning light, look like finger boards pointing to the Creator and giver of all that is good to mankind."

It was a well-afforded appreciation of the land. The narrow gauge ran over the hills and into the northern valleys of the field as well:

The Bradford Oil Basin

"The road at this point is supposed to be about four hundred and sixty feet higher than at Kendall, and it is fully two hundred feet higher than at the level of Boyd Valley. After leaving the summit the road runs curving in and out along the side of the mountain through deep, smooth cuts and over curved, high trestle works, gradually dropping lower and lower into Knapp's Creek valley until Judson's Camp is passed, the cars stop, and Rixford appears on our right."

"From Rixford to Eldred the road runs along the valley, through a beautiful area interspersed with stumps and swamps; here and there a dwelling or a sawmill... "

The wilderness was easier to enjoy by train, with the rails to ride on in comfort, and with civilization at the track's end at Bradford.

The Oil Towns

The small towns of the oil district flourished as long as their derricks continued to produce. Many achieved enough size and importance that they still exist today, mostly as quiet, rural villages.

Many, such as Barnum and Indian Creek remain as place names only. Some, such as McKean's Mills, have disappeared altogether. Perhaps the shortest-lived oil town of all was Jo Jo in the south of the county. It sprang into existence in the winter of 1885-1886, and was practically deserted by April of 1886.

One of the major towns was Allegheny Bridge, or Eldred. Eldred was strategic because it was the site of the main bridge over the Allegheny River and into the oil fields from the North. It was not itself primarily an oil producing town, but was built at the edge of the oil fields and provided a base for workers at Haymaker and Indian Creek. It was also home to the Eldred Refinery, later producing Betty Blue Gasoline.

Jon C. Cawley

Post card views of Eldred and Farmers Valley.

The Bradford Oil Basin

THE COURT HOUSE, SMETHPORT, PA.

Section of Main Street, Smethport, Pa.

Jon C. Cawley

Eldred was located on the Eldred-Olean road, and was soon the terminus of the Kendall and Eldred Railroad. Later the Bradford, Eldred, and Cuba railroad and the Bradford, Bordell, and Kinzua road were built through the town as well.

The town was incorporated as Eldred in 1880. It had its own barrel coopers, glass plants, and clay and shale pits. Some of the more important nitroglycerine factories for well shooting were located here as well (Later during World War II, these made ammunition for the war effort.). Eldred had a rather famous wooden handle and baseball bat factory, and a newspaper, the Eldred Eagle, established in 1878.

Nearby was Smethport, the county seat, named after a Dutch banker who had invested in early settlements at Ceres. Smethport became the county seat in 1826. It has had three county courthouses. The first was built in 1825-1826, and contained sheriff's quarters, a courtroom, and two cells for prisoners. This was torn down in 1850, and a second building erected in 1851. This building served until it burned to the ground. A third courthouse was built in 1881. The town was incorporated in 1853. Its major industries (excepting oil) have been glass, chemicals, and agriculture. The Mckean County Miner was established in 1862 and the McKean Democrat in 1874.

Other towns included Indian Creek and Haymaker, with their long wooden sidewalks, their hotels, boarding houses, and saloons. Rixford was thought by some to be the richest field of the oil region in 1878. Oil towns were tallied in and placed on the local map as they were noticed. Duke Center was welcomed in January of 1878:

"**The village Duke Center now in view of its new buildings and muddy streets is entitled to its place among the oil towns of McKean County.**" (The Bradford Era, January 1878)

The Bradford Oil Basin

Jon C. Cawley

Oil Derricks at Duke Center from a stereo-card.

The Bradford Oil Basin

Duke Center, bypassed by the railroads, built a wooden plank road all the way to Eldred, a distance of about ten miles.

Towns like Marion City grew quickly out of nothing:

"Marion City is becoming a place of some importance as an oil town in the woods at the second trestle. There is but one house, but there will be others built as soon as lumber can be obtained. There are eight producing wells, and about that many drilling." (The Bradford Era, November 1878)

The towns and small cities continued to grow and expand. Eventually there were over one hundred sixty of them in McKean County alone. The oil region was well established and the Age of Bradford begun.

BR-50 Bird's-Eye View of Main Street and Business Section, Bradford, Pa.

The Bradford Oil Basin

Jon C. Cawley

Chapter 2
A Short History of Bradford

At the heart of the northern oil region is Bradford City. Hidden among the McKean County forests, Bradford has had a reputation as the "Oildorado", or drillers' paradise, of Pennsylvania. After the initial discoveries, drillers from the southern fields rushed to the newfound "oildom to the north". The green oil of the Bradford sand seemed a sure thing. Wells put down in the region almost always produced. For many years, the few dry holes that did show up usually demarked the edges of the great Pool.

As the oil trade grew, Bradford was bought and built with the flow of the local crude oil. Soon there were railroads and refineries, pipelines and oil field supply companies, theatres and fine arts guilds. By the 1880s, Bradford was officially known as "The High Grade Oil Metropolis of the World."

Early Bradford

In the early 1800s, "Bradford" was nothing more than cluster of dwellings on the banks of the Tunungwant (or Tuna) Creek. Like many small settlements of the north Pennsylvania wilds, it was based mainly on the lumbering of the great forest stands, primarily white pine. The lower region of the Tunungwant Creek had several advantages to lumbermen.

The Bradford Oil Basin

The surrounding hills made the valley a great natural amphitheater. The forest here was of good quality, and it was easily cut from the rolling slopes above the creek. There was water power as needed for local sawmills, and the nearby Allegheny River system gave natural transportation to markets at Pittsburgh, Wheeling, Cincinnati, and Louisville, even as far as New Orleans.

In 1827, a sawmill owner, John F. Melvin, and fourteen other residents of the valley petitioned for the formation of Bradford Township. This marked the Bradford region as a separate unit from surrounding townships.

By 1937, Col. Levitt C. Little had joined the small community on the creek banks. He took an active citizenship and the name Littleton was adopted for the town. With time however, the place-name Littleton was dropped in favor of Bradford as the official title. Bradford was the name of the local post office after 1854.

Population in the Bradford region grew slowly in the middle 1800s. By 1860 there were only about 8800 people in all of McKean County. The area was largely bypassed by western migration; the Allegheny Portage route was less popular after the completion of the Erie Canal to the north in 1826. (The 1860 population of Cattaraugus County, New York, closer to the canal route was 43,800 people.)

In 1872, Bradford was incorporated as a borough. Its population was still very small. As late as 1876, the year that the Bradford oil boom really began, there were only about four hundred people in Bradford, and less than 1750 in the whole Tunungwant Valley. The town at this time was still a rough-hewn cluster of buildings. Most well recognized were the log schoolhouse built in 1835, the Bradford House Hotel built in 1838, and the Old Red Store, a clay-painted building built in 1853 that served both as post office and a place of busi-

Oil Wells on Allegheny River, Bradford, Pa.

Early views along the Allegheny River.

341 A Good Producing Oil Field, Bradford, Pa

ness. The town had its first newspaper, the Bradford Miner, by 1858. By 1876, Bradford had also claimed the machine shops of a lumber railroad.

By 1871, the first cautious drilling was taking place around Bradford. Drilling started north in the New York State end of the Bradford field. Drillers kept moving up the creek from Limestone, and by 1875, Bradford and nearby Tarport were being drilled. Initially, when no oil was found at the shallow depths of the Venango sands, many thought the area around Bradford to be barren of oil. However, as the Bradford third sand was struck at nearly twice the depth of the Venango, the oil flowed, and more derricks were quickly built. The largest wells- the Crocker well on the Kingsbury farm, the Olmstead well on the Crooks farm, and the Lewis Emery well at South Bradford sparked the boom that was to follow.

The Bradford Boom Years 1875-1881

The time between 1875 and 1881 was a time of phenomenal growth for Bradford. As the oil derricks began being raised on the banks of Kendall Creek, Foster Brook, and the Tunungwant, Bradford began its rise to oil fame.

"Twenty years of indescribable excitement followed the discovery of oil. The valley swarmed with prospectors. The streets were ground into mire by heavily loaded teams. The strife to secure leases of promising territory was keen and many farmers thinking to take advantage of the FarmPetition made haste to sell royalty interests." (Stone 1926)

The growing community was besieged with drillers and speculators from the southern fields. Many veterans of the Civil War were seeking the oil fields as well, and numbers of them came to settle with the new boom. Within a year or two, the population of Bradford had

The Bradford Oil Basin

more than tripled. Bradford remained as the center of the oil rush. As the derricks grew in number, the region was filled with smaller independent oil towns that looked upon Bradford as their capital metropolis:

During 1876, the vicinity of Tarport, now East Bradford, at the mouth of the Kendall Creek was the center of excitement, with drilling extending some distance up West Branch and up East Branch as far as Degolia. Early in 1877, interest shifted to Foster Brook. Some production had also been encountered by this time at Big Shanty, seven miles south of Bradford and in the vicinity of Rock City and Four Mile, ten miles northeast of Bradford.

During the spring of that year, gas was found along Indian Creek but it was not until the latter part of the summer that the oil pay beneath the gas was discovered. During June 1877, Kendall Creek became the Mecca of the oilmen and the Knapp's Creek excitement started during August of that year. September saw the opening of the Quintuple tract southwest of Bradford between the East and West Branches of the Tunungwant. (Fettke 1938).

By the 1870s, oil pipelines were being laid from the Tuna Valley. The region was over-producing, and storage and transportation were in short supply. During this oil flood, the oil market---controlled in part by outside interests and set by Bradford production -- was often unstable.

With the towns and oil excitement, quickly came the railroads. As early as 1856, Bradford was home to a branch line of the newly formed Erie system. By the 1870s, there was a great competing network of both narrow and standard gauge railroads throughout the Bradford region. Railroads included the Kendall and Eldred, the Olean, Bradford and Warren, the Big Level and Kinzua, and the Bradford, Bordell, and Kinzua, among others.

Bradford, Pa. Main St.

MAIN STREET, BRADFORD, PA., VIEW FROM ST. JAMES HOTEL.

The Bradford Oil Basin

Peg Leg Railroad. Quit Business 1880. Bradford, Pa.

Jon C. Cawley

Perhaps the most interesting of the Bradford region lines was the Bradford and Foster Brook Railroad that ran in the Tuna Valley from 1877 to 1879. This road, better known as the "Pegleg Railroad," was a steam-driven monorail line, and is generally considered to have been the first successful monorail system in the world. The train rode on a heavy wooden "sleeper" rail, which was mounted on posts. The LaFrance Manufacturing Company of Elmira, New York built the Pegleg engines. Fare on the four-mile line was five cents a mile, and by May of 1878, the train's daily income was one hundred dollars.

The Pegleg ran successfully until January of 1879, when a boiler explosion destroyed an engine and killed six men. This was enough to bring an untimely end to the small enterprise; it was sold at a sheriff's sale in February and was dismantled soon thereafter.

Bradford railroads were well known for the way they managed the region's rugged terrain. Roads ran from flat, prairie levels to grades of 264 feet or more per mile. Grades that could not be made were cut through, built around, or bridged over. Where deep valley ravines cut the Big Level region, the challenge was greatest for bridging, and this was the home of the famous Kinzua Viaduct.

In 1880, General Thomas L. Kane was attempting to extend the Erie Railroad lines southward from Bradford. Kane's New York, Lake Erie, and Western Railroad and Coal Company was quickly faced with the Kinzua Valley, more than two thousand feet from valley side to valley side at grade level. An offer of bridging came from Anthony Bonzano, bridge engineer of the Clarke Reeves Division of the Phoenix Bridge Company. Bonzano said he would build an iron bridge "a thousand feet high" if the money was provided. Money was not a problem in the region, and the great bridge was built in 1882.

Kinzua Bridge, Bradford, Pa.
Height 301 feet, Length 2100 feet.

Jon C. Cawley

The Bradford Oil Basin

The viaduct was built in eighty days. It was 205 feet long and 301 feet high at center, winning it distinction as the highest railroad bridge in the world. The bolted ironwork weighed over three million pounds. The bridge was supported by twenty great towers, each of which was built by gin pole from the one before, without the aid of scaffolding.

When completed, the bridge was judged by many as "one of the wonders of the modern world." The original Kinzua Viaduct was dismantled and rebuilt with steel in 1900, to accommodate the heavier locomotives of the later years. This second structure remained in use until 1959, and still stands in the forested Kinzua Valley south of Bradford City.

In 1879, Bradford officially became a city. The change came as oil discoveries around Tuna Creek were at their greatest, and the hillsides around the new city were lit day and night by torches of natural gas from the wells. Bradford was the hot spot of the era. Still it was an oil city---people coming to the region as late as 1882 told of being stuck in the mud of the unpaved city streets.

On the east side of Bradford was Tarport, a rough and tumble section of town. Bradford region had its share of saloons and gambling of loose ladies (referred to as "nymphs du pave"), and houses of ill repute. As Bradford itself became more civilized, many of the coarser businesses moved as far a Tarport, which quickly gained a reputation reminiscent of Pithole's some years earlier.

However, Bradford was unlike the "oil boom towns" that had come before. "Bradford oil was sound"----and local producers and refiners quickly developed strong loyalties to their business and to Bradford. Law came quickly, and the city began a period of municipal growth and improvement. Streets were paved, volunteer fire companies were housed, and a city hall was built.

The Bradford Oil Basin

Congress Street, Bradford, Pa.

Main St., Looking West, Bradford, Pa.

The establishment of public water and sewage companies quickly followed. Natural gas was piped in to light the streets, and the methodical supply system (begun by Charles E. Hequembourg) is considered to have been the first of its kind. Schools were erected, a library was established, and by 1877, the first of Bradford's world-known oil exchanges had been built. News of this growth and proliferation was duly recorded by the city's newspaper, The Bradford Era:

"The New Era had been founded back in 1875 to judge whether the McKean field was an "oil excitement" or an "oil development." By the end of 1881, it was clear to all that the Bradford field had completely changed the oil picture; that the oilman's horizon had been infinitely widened, and that it was capable of indefinite expansion during the next fifty years." (Lawrence 1938)

Bradford After 1882

After the oil boom years of the 19th century, Bradford production slipped into a phase of decline. As gas pressure in the fields lessened, pumping became the rule, and finally local producers turned to water flooding in the region. Bradford of the time was stable. Belief in the oil and the future of the city counted for much: non-oil businesses flourished. There was a hospital and a children's home, and the streets carried trolleys and automobiles by the 1920s." In 1923, Kendall introduced its first line of low temperature motor oils, and began a new era of its own. Their new refinery was completed in the 1930s.

By the 1930s, the ingenuity and effort of Bradford producers had resurrected the now marginal oil fields. With successful water flooding, there was a second upward trend in region oil production. This peaked at a second, seemingly impossible high of 16 million barrels in 1937. Bradford saw a short secondary influx of oil money and prestige.

The Bradford Oil Basin

This boom of the 1930s protected the Bradford region somewhat from the trying times of the Great Depression. By the 1950s, however, the oil output was again beginning to slip.

As oil began to play a lessening role, Bradford wisely offered itself as home to manufacturing and service industries as well. Names like Corning (electronics), Zippo (lighters), Owens Illinois (paper containers), and Case (cutlery) have joined the more traditional oil-related names like Kendall and Quaker State, Dresser and Bovaird The forests once again had leading potential, and lumbering companies also took new interest in the area. By 1970, the city still had over 12,000 people and boasted a branch campus of the University of Pittsburgh.

Bradford Since 1980

By 1986, however, Bradford had little remaining of its fame and glory from the "oil boom days". With decreased oil production, came the "(economic) crash after the boom". Bradford was still declining after its 1937 flooding peak. Despite strong drilling trends in the 1970s, oil production of the basin was again falling to "bare subsistence" levels. From six million barrels in 1960, production had fallen to about three million barrels in 1980. Bradford's era of aqueous extraction (water flooding) was now coming to an end, while tertiary recovery (the supposed "next hope") has still been a questionable prospect. Now both production and industry in the Bradford basin were again changing profoundly, with the decline of the green-gold economic base of oil in the region.

Bradford itself showed serious wear, as the oil production continued to slow. The oil fields had seen their share of miracles, and Bradford had long since lost the bustling, prosperous economy that it had at the beginning of the Century. Its streets and roads were still narrow and steep, and tended to be unkempt; access to the

S. R. DRESSER RESIDENCE. - BRADFORD, PA.

The Bradford Oil Basin

outside world remained relatively limited. Sewage and maintenance systems had become outdated, and there were problems with the local water as well. With lessening income, many residents of Bradford found local costs of living and taxes oppressive if not unmanageable; houses and properties were being sold, and programs of low-income government housing appeared to be adding to the local erosion.

Despite an optimistic outlook voiced by the McKean County Planning Commission, in the late 1970s, Bradford lost one quarter of its population and business from 1960 to 1980. From a population 15,061 in 1960, Bradford had only 12,672 in 1970, and 11,211 in 1980. Productive Bradfordians continued to leave or age, and young people who left for college seldom returned.

At present, there is still little opportunity in the area for many of them. Local investment was down, and a sort of survival ethic had slowly begun to take its place. Things were extremely tight in many ways, and the future of Bradford itself might have been ultimately in question. The population of Bradford in 1990 was approximately 9,625. The approximate number of families at that time was estimated at about 4,477.

Nevertheless, it is a tribute to the region that, after a hundred years, the city still exists. For an "oil boom town", Bradford has enjoyed an incredibly long and stable existence. Even now, Bradford retains much of the glory of its gilded past. Its architecture hails from richer years; it is a town of Victorian cornices and Hanley brick. On its streets, small modern houses stand between Victorian mansions. Bradford industry ranges from antique to high technology.

Even Bradford's social customs, a strange mixture of fine taste and roughness, persist. The old names are still well remembered: Emery, Koch, Blaisdell, Dresser, and Dorn, are still well known.

Even now, in Bradford, the excitement of oil still lingers, rather, is alive and well. There are still oil leases in the woods of McKean County. The refineries are at the center of town; great tanks are scattered on the highlands, and along the Tunungwant Creek.

There are wells in backyards and in front lawns, in the parking lots of the business district, and among the quiet graves of the local cemeteries. Pumpjacks, engine houses, and multiple pump rod lines still dominate the residential areas. Even today, the dreams of Bradford are based in and on the hope of, and quality of, its green crude oil. It is a history and a dream worth preserving.

31. L. A. EMERY RESIDENCE, BRADFORD, PA.

BRADFORD, PA.

Jon C. Cawley

Chapter 3
The Era of Derricks and Drilling

Drilling in the Bradford Field

The first oil wells in Bradford were wildcats; they were drilled without much geological knowledge of the area. No major sands outcropped in the Bradford basin, and there was little geological information to lead the drillers. Bradford was a tiny village of less than four hundred people in 1858, and the earliest wells were literally drilled in wilderness.

In 1861 several Bradford citizens spring poled a well where West Corydon Street is today----the well was two hundred feet deep, and was later deepened to eight hundred seventy five feet. This well missed the oil sands by about one hundred and fifty feet, and the hole was abandoned. In 1862, the Barnsdall Company bored to about eight hundred seventy five feet and also got nothing. North in Allegany, New York, Harvey Pierce and Dean Goodrich spring poled to five hundred and seventy feet. In Custer City in 1865, a well was sunk to nine hundred feet----within two hundred and thirty feet of the Bradford Sand there----and was also abandoned.

Although these first wells were not successful commercially, they were important technologically; they were all kicked down with salt drillers' spring pole systems. We must realize the huge effort put forth to kick down a well by hand to hundreds of feet. The spring pole system consisted of a sapling bent over the area where the well was to be drilled.

A crib or hollow log casing was sunk to bedrock, and the bit string was then attached to the springy end of the sapling by a loop. Rope was supplied from another wooden mast to the side. The auger or bit was made of cast iron and was dressed in a V-shape. It was about an inch and a half wide and about three and a half feet long.

This point was mounted on heavy ash wood poles, which were connected to form a heavy and formidable drill string.4 A foot strap or treadle on the sapling above allowed the driller to stomp the bit string vigorously against the bottom of the hole. And the chisel-like bit was given a circular motion by the second driller, who turned the wooden poles as the string was jerked down. As the chisel end pulverized the rock to sand, the poles were drawn out periodically, and a sand pump was lowered into the hole. The auger was taken from the pole and a copper tube with catch valves in the bottom end was dropped down to clear out the sand.

An experienced driller could drill perhaps three feet per day (Experience came fast once one's legs got in shape). And although they didn't find oil in the Bradford district, theirs was still an admirable endeavor, especially where they drilled through the Knapp Creek Sandstone or the Olean Conglomerate, the most resistant layers in the Bradford field.

The first paying wells came with more sophisticated drilling methods. In 1865, a Yankee businessman by the name of Job Moses, came to Allegany, and following the belief that oil flowed beneath stream courses, set up a derrick and steam engine to drill. The derrick for Moses' Discovery well stood sixty-two feet high, and was shingled with matched one-inch boards. The crown pulley at the derrick top had a shingled roof, the rig contained a complete blacksmith set, and the chimney of

Spring-pole drilling system.

the forge was built with red brick. (This impressive building stood until 1882, when it was destroyed by a freak tornado----it was not rebuilt.) The Moses well was drilled to about 1119 feet, and in November it "came in" at ten barrels of oil per day.

In 1871, the Foster Oil Company, supported by Job Moses, drilled on the Gilbert farm in Bradford. This well also struck the Bradford Third, and came in at about ten barrels per day. By August of 1875 there were six producing wells in the Bradford Basin.

On October fifth of 1875, however, the Crocker welllo~ the Kingsbury farm at Bradford came in with a gusher at 300 barrels a day, and the true strength of the Bradford Third Sand was realized. Bradford grade crude imir~eldiately brought $2.00 a barrel at market (opposed to $1.25 for other oil), and Bradford was opened to the world.

Knowledge, intent, and improved drilling methods put the prodigious Bradford Third Sand within reach of the drill bit. And after 1875 production skyrocketed in the Bradford basin; 25000 barrels in 1875, six million in 1878, twenty-two million in 1881.1 In those years, people flooded into the region from the southern fields, and hundreds of oil towns sprang up in the four county area. Duke Center, Marion City, Rixford, Tarport, Cyclone, Barnum, and Eldred were points of interest and towns known by the world. As the boom towns grew, railroads were built, and the forests were felled and re-erected as the derricks were built to tap the green oil of the Bradford Third.

During the boom years after 1875, the Standard percussion rig was the best-known and recognized drilling system in the Bradford field. The Standard derrick was seventy-two feet high, and was built of massive beams cut from the trees (usually chestnut) of the high ridgetops.

The Bradford Oil Basin

In the Standard system, the string of drilling tools was suspended from a massive Nuecomen-like walking beam, which, via steam or gas power, supplied the up, and down motion for boring the hole. The manila rope or cable, which supported the drilling tools, was wound on a great powered bull wheel from which it could be taken in or let out. This rope was then looped over pulleys in the crown block at the top of the derrick, and attached by means of a temper screw to the derrick end of the walking beam. The temper screw was an ingenious invention; a simple clutch that adjusted the cable to the depth of of the hole allowed the string to be turned from above and the cable be attached or detached from the walking beam power at will.

To drill, a casing pipe was first driven to bedrock within the base of the derrick. The tool string was then "spudded in"---that is, the walking beam was disconnected from the drawn up cable. A line from the band wheel power was attached directly to the cable above the bull wheel---the turning of the band wheel imparted horizontal motion to the bull wheel cable, which in turn translated itself over the crown block pulleys, and moved the drilling tools vertically. This smaller vertical motion allowed for surface work to start a drill hole.

Once the hole was deep enough to accept the drill string, the drillers drove a casing pipe into the well to protect the well from surface water and cave-ins. The well was cased to about three hundred fifty feet and was sealed from outside water by leather bags of flax seed, which swelled and filled the space between the casing and the rock walls of the hole. Later, rubber packers, invented by Bradford's S. L. Dresser replaced these flax seed bags.

Drilling then proceeded by walking beam power down through the rock. As the temper screw was let out to its full length of about five feet, the drill bit pulverized the rock in the hole. The drill string was drawn

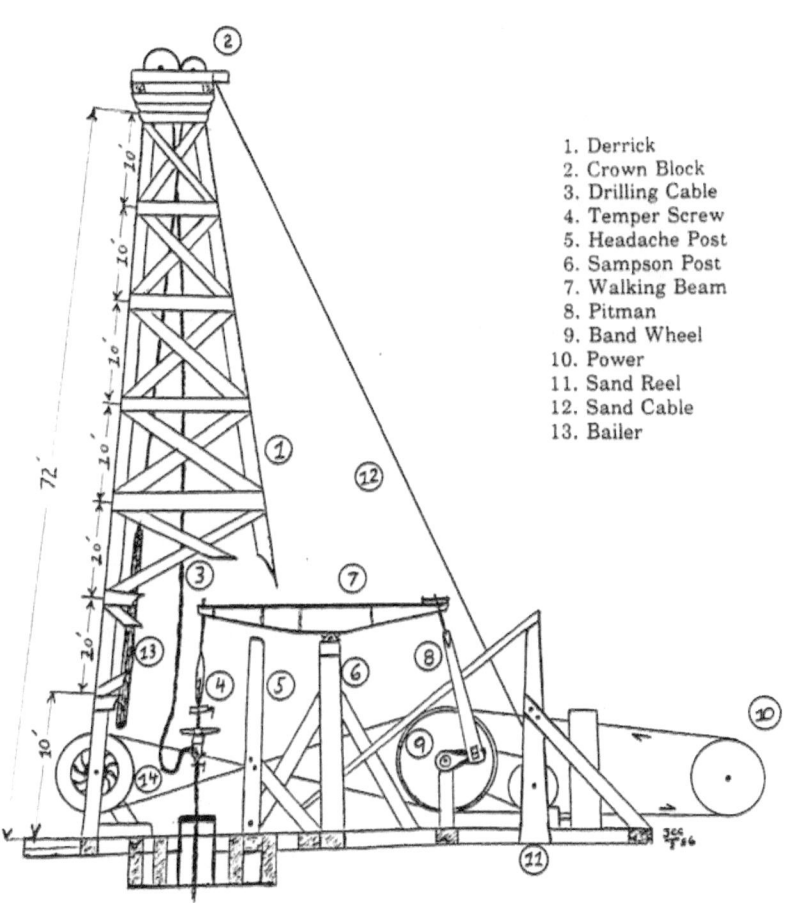

1. Derrick
2. Crown Block
3. Drilling Cable
4. Temper Screw
5. Headache Post
6. Sampson Post
7. Walking Beam
8. Pitman
9. Band Wheel
10. Power
11. Sand Reel
12. Sand Cable
13. Bailer

Standard drilling rig.

The Bradford Oil Basin

In the heart of the Oil Country: Scene near Bradford, Pa.

up periodically, and a sand pump or bailer was lowered to clean out the hole. This change in operation allowed the drilling tools to be changed and dull drill bits to be sharpened.

This was done at the derrick's forge; the drill bit was heated to "cherry red" (actually about white hot), and was hammered to a new point. Both the driller and his tool dresser stood over the glowing metal and hammered it from two sides to give it the correct shape.

To save time in such work, the bailer or sand pump was strung on separate rigging within the derrick. In this way, the hole could be cleaned as the drill string was being worked on The sand cable ran from a powered sand reel, over the crown block, and down to the well hole. When not in use, the bailer and cable were pulled to one side inside the derrick.

The drilling and cleaning cycles went on until the hole was complete---the sand was reached, or the hole was declared dry. The process went slowly, with drillers making six to eight feet per day. Thus, even working continuous twelve-hour shifts, drilling a well could still take several weeks or months.

As drilling systems became more complex, so did the drill string. No longer were the "pocket weight" bits of the spring pole drillers useful. By 1873 drilling bits were made of Norway steel, and weighed four hundred pounds. They were drawn to a width of five and a half inches, and came in many shapes and forms.1~ And with advanced drill bits came other tools; reamers, special bailers, and corers for various needs and occasions. Early on came the jars. Invented by William Morris in 1831, this set of links in the drill string absorbed much of the percussion force of the drilling.

The jars allowed for vertical movement within the string components in the drill hole, and were invaluable when the drill string became stuck or wedged while drilling. By 1875 it was not unusual for the

The Bradford Oil Basin

manila drilling rope to support thirty feet of tool string within the hole. And because drill strings tended to break or become unscrewed in the hole, a full set of various fishing tools-grapplers, clutches and hooks-were available to retrieve them. Such tools were standard equipment for any drilling operation until well into the 20th century, and many are still in use today.

The Standard cable derrick was most common, but not alone in the Bradford field. The California cable rigs, for example, boasted a larger derrick (106 feet) and the addition of a secondary calf wheel to aid in swinging large strings of tools. Many of the rigs in the Bradford Basin were actually hybrid combinations of Standard and California rigs. "What worked" was usually the only rule in the Bradford operations.

A more interesting diversion from American cable systems was the Canadian pole drilling system. Adapted for Canadian fields, this rig retained the ash drilling rods of the old spring pole system combined with the drill strings and derricks of the cable rigs. The Canadian drilling rods were made of long grained ash, and were about thirty-seven feet long. These were screwed together with iron couplings, and acted like the weights and drilling cable of other rigs.

The bit used was a light auger type weighing about sixty pounds, and about one and a half feet in length. Like the cable systems, the Canadian rig used a walking beam, but there was no temper screw or bull wheel. Instead, a ratchet-and-pawl arrangement known as a "slipper out," worked by a hand line to the driller, allowed the drilling string to drop slowly. The Canadian system was practical in clay regions like Pennsylvania, but was never as popular as the Standard cable rigs were.

After the turn of the Century, many oil contractors began to look for cheaper methods of drilling wells

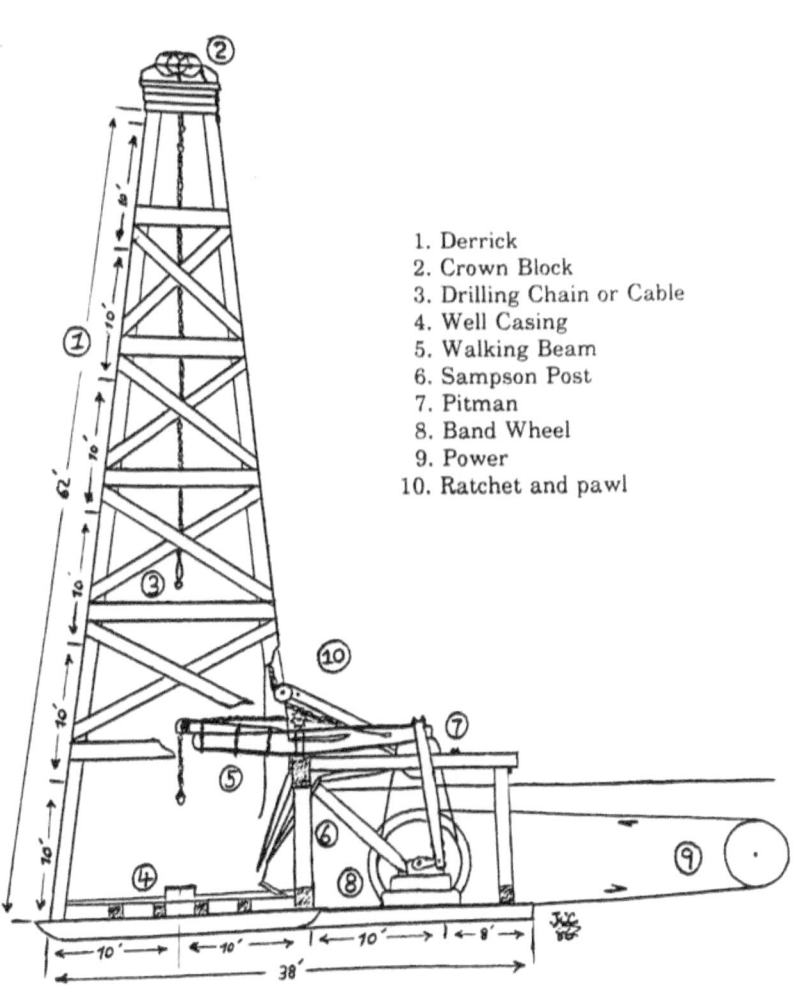

Canadian pole-drilling system.

The Bradford Oil Basin

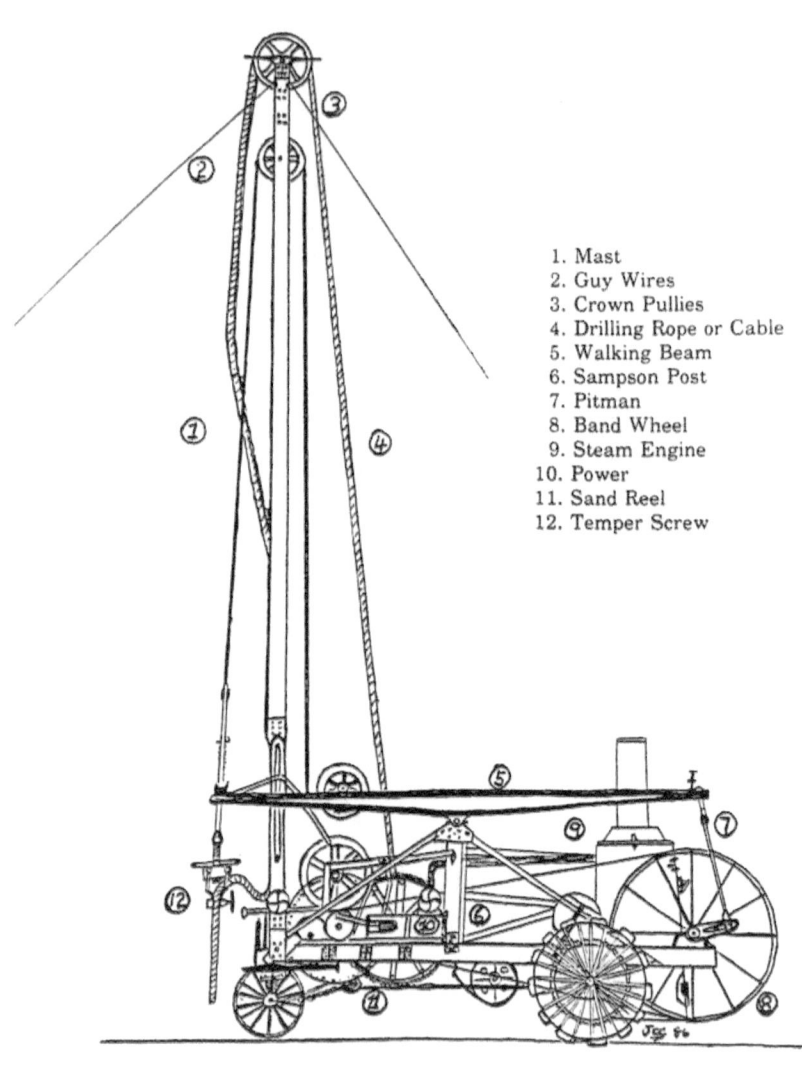

1. Mast
2. Guy Wires
3. Crown Pullies
4. Drilling Rope or Cable
5. Walking Beam
6. Sampson Post
7. Pitman
8. Band Wheel
9. Steam Engine
10. Power
11. Sand Reel
12. Temper Screw

Early portable drilling machine.

in the Bradford fields. Timber for derricks was becoming scarce, and Bradfordians did not particularly like the newer steel derricks (Metal derricks, it seemed, did not creak and groan under undue pressure or strain like wood. Instead, when stressed, they simply collapsed quietly and without warning.)

As a result, many types of portable drilling and spudding machines came into use. For the relatively shallow and simple drilling around Bradford, portable machines made sense. The major limitations of such machines lay primarily with the technology of the times. Although many portable rigs worked in theory, a good number of them never worked well in practice. Portable machines replaced the traditional derrick with a single mast supported by guy wires. The early machines, such as the Columbia driller were operated by their own steam power.

The machines were lightweight with a much reduced bull wheel and a friction controlled sand wheel. The band wheel shaft carried smaller, lighter cranks and pitman, which operated the walking beam. As may be expected, even the best of the earlier machines were successful only for shallower holes to about one thousand feet.

By the 1940s however, several Bradford companies introduced more advanced portable spudding machines. These machines were more surely built, and used diesel and gasoline engines. They were in the long run less costly, and were useful in reconditioning older wells in the fields as well as drilling new ones. With time, the spudding machines have entirely replaced the oil derricks in and around Bradford.

For a short time after the major petroleum labs were established at Penn State (1929) and Bradford (1943), the need for diamond drill cores forced rotary drilling into the Bradford Basin.

The Bradford Oil Basin

Rotary systems were used in conjunction with cable wells at Bradford---the well was drilled with cable tools to the depth of the sands to be cored. A crew of outside specialists then assembled the rotary system, and the sands were cut. Any further drilling was then done by cable.

In the rotary system drill pipe was attached to swivel within the derrick, and a rotary diamond bit was lowered at the end of the drill pipe into the hole. Cores were taken as the pipe was rotated within the well by a powered rotary table on the floor of the derrick. Mud was pumped from a "slush pit" down through the center of the drill pipe to cool and lubricate the drill bit and the walls of the pipe.

Although still used elsewhere, rotary drilling was expensive, and Bradford producers were generally unaccustomed to it. When cable type core tools were made available by Bovaird Company and others, rotary drilling saw little further application in the Bradford field.

Thus overall, for the greater part of a century after 1865, the oil derrick reigned supreme in the Bradford fields. In that time 40000 holes were recorded, with perhaps that many again that went unreported The chuffing of the steam engines, the sight of the derricks standing above the tree tops; the clouds of white steam, and the oily smell of the creek bottoms were symbols of progress and prosperity to the people of Bradford.

Children grew up climbing derricks when they could get away with it. They stood among the crown blocks, and rode the walking beams. and when they were grown they became the drillers and producers; to learn the language of the temper screw, and the sound of drill tools biting at the bottom of the hole.

Drilling was the legacy of the Bradford Basin.

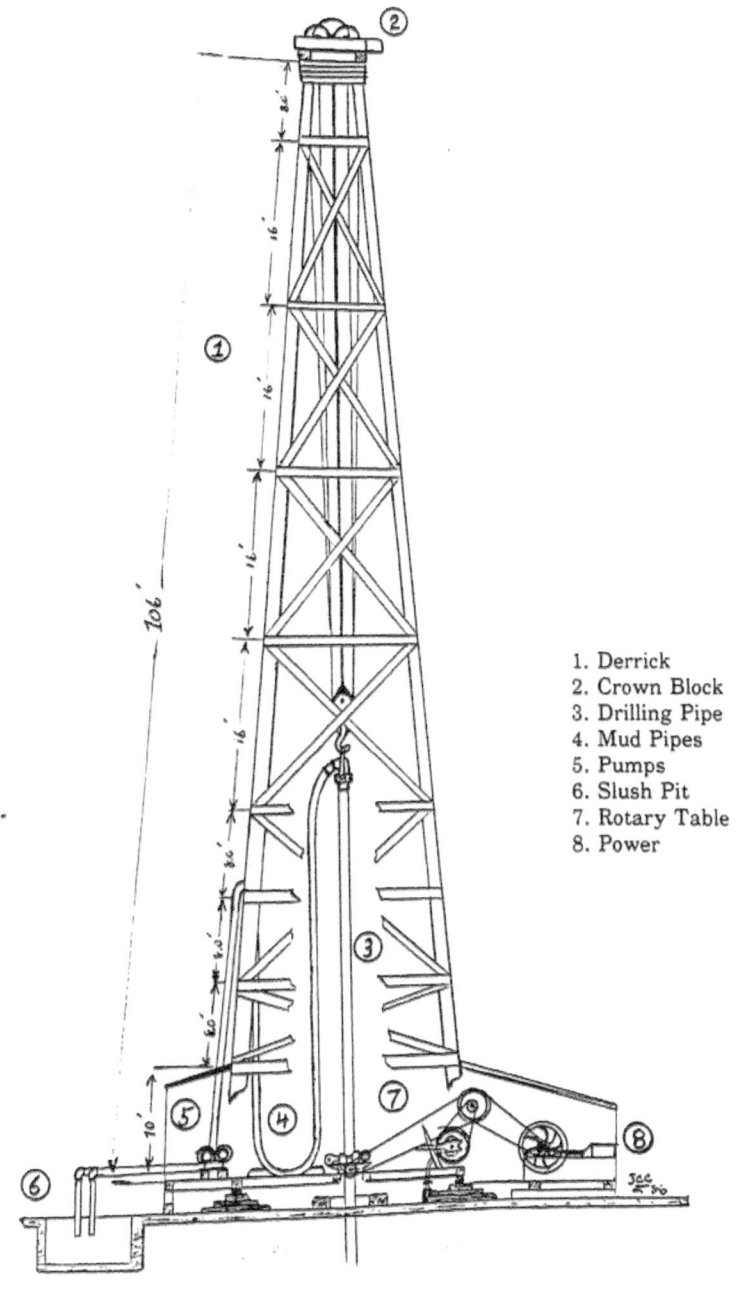

Rotary drilling rig.

Chapter 4
Bringing in the Wells

After a hundred years of drilling, no-one yet knows exactly how percussion tools act in a drill hole. The physics of two thousand pounds of cutting weight jerked up and down at the end of a cable in a drill hole can be awfully complex.

Traditionally, for example, it was a driller's job to constantly turn the temper screw out slowly, both to keep the bit in contact with the bottom of the hole, and also to give circular motion to the bit. This was thought essential to keep the hole round and straight. In actuality, when the bit is jerked up and down, the drilling rope expands and contracts, and this action alone spins the tools at the bottom of the hole. The tireless hours that the driller spent turning the temper screw in and out probably didn't have much of an effect on keeping the hole round.

Time at the temper screw, however, was not lost, for no matter what the drill string was actually doing below, the vibrations of the bit cutting through the rock were transmitted upward, and could be felt at the temper screw. An experienced driller who knew the feel of the screw could pretty well tell what sort of rock the bit was cutting through at any time. This knowledge came by matching the feel of the temper screw with the rock fragments brought up by the bailer periodically. After awhile a driller could feel his way down through the rock layers with good efficiency.

The Bradford Oil Basin

When the feel of the temper screw told of soft shales, he knew that he could drill more quickly. When the rock layers were hard and resistant, he would back off and use lighter strokes to avoid breaking the bit or drill string. When the bit was dull, or the hole needed to be bailed, he could feel it.

By and by the drill would bounce on the hard "shell" layer that capped the oil-bearing sand. When the driller believed that he was near a sand, the open fires of the steam boiler, the forge, and the yellow dog lamps were quickly extinguished. Once the sand was reached, a gusher or a rush of natural gas could mean instant death if it caught fire. This was a dangerous point, and many men lost their lives when they hit gas unexpectedly, or when their derricks caught fire, claiming the single well or raging through the whole of the surrounding field.

Often, a "snuff line" was laid from the steam boiler to the mouth of the well. When a volume of gas was struck, the high pressure steam was turned on in hopes that it would mix with the gas and help disperse it. Whether this safety feature helped much or not is questionable.

After the sand was reached, the oil sometimes spurted immediately from the ground under its own massive pressure in a huge fountain. Gushers were smaller in the Bradford field because the gas pressure was a bit less (estimated at 350 pounds, original). But they were numerous in the early years, and when a well came in as a gusher people in the field put up their umbrellas and celebrated.

If no gusher occurred, however, drilling continued slowly into the sand, and the hole was bailed often for signs of oil. Good oil sand was chocolate brown in color--a strikingly handsome rock containing a mother-of-pearl of fossil shell against dark background. It was a favorite driller's test to pour hot water from the steam

Jon C. Cawley

The Bradford Oil Basin

A GUSHER IN THE OIL REGION OF PENNSYLVANIA
PICTURESQUE BALTIMORE & OHIO RAILROAD

Above:
Bovaird
Yellow Dog
Derrick Lamp.

Below:
Water flowing
from an
exploded well
at
Indian Creek.

The Bradford Oil Basin

engine over the rock fragments in the bailer and look for the rainbows that indicated that oil was present. Lighter colored sands often indicated water in in the sand layer, and were taken as a bad sign.

Because Pennsylvania sands have small pore spaces, the oil was often sluggish even when it was present; the oil filled the hole slowly, and the production was limited. In the 1860s producers discovered that setting off explosives inside the oil wells increased production greatly by fracturing the sand and increasing the surface area of the hole. The force of the explosion formed a cavity in the oil layer in which the oil could better collect and flow.

"Shooting" oil wells was one of the most impressive spectacles of the oil regions. Because dynamite was not always effective or dependable at the bottom of a well hole, tin shells of nitroglycerine were used instead. For shooting, the well was cleaned out, and a shooting specialist was called in to the well. The tin shells, or torpedos, were positioned over the well on rigging, and were filled with about sixty quarts of nitroglycerine each from small metal cans. A fulminate of mercury fuse capped the top shell, and the string was lowered into the well hole. It was important that the shells be centered well within the sand layer so that the explosive force would do the most good. A column of oil or water was usually added to tamp down the charge. When all was in place, the shooter would drop a "go-devil" into the borehole.

The go-devil was a heavy (about eight pound~ piece of solid cast iron fitted with wings to make it fail straight down the hole. The go-devil would strike the detonator squarely and with great force, exploding the shells. Eventually an automated fuse, or squib, replaced the falling weight, but for a long time, the go-devil was almost mythical as a symbol of simplicity and effectiveness in the oil fields.

The Bradford Oil Basin

After dropping the weight, the shooter would stand by the drill hole until the shot was exploded. This was signalled by a violent tremor in the casing pipe of the well. He would then run to a safe distance and wait for the results. It took several seconds for the force of the explosion to reach the top of the drill hole. After that short infinity of time, the ground began to rumble with increasing intensity. There was a secondary clatter of material hitting the inside of the derrick, then with a great roar, a huge main cloud of water, sand, oil, and debris was hurled upward from the mouth of the hole. The effect was of a brief, violent, strangely colored gusher which obscured the derrick and sent a rain of oily sand over everything nearby.

A secondary spectacle came soon after, when the empty nitroglycerine cans---which were considered to be too dangerous to be salvaged-were exploded. As an observer in 1914 described:

"A tin is never used twice for loading nitro-glycerine, nor, indeed for any other purpose; the risk is too great. No matter how carefully and apparently completely a tin may be emptied, tiny beads or a thin film of the liquid are certain to adhere to the interior sides of the cans; consequently, when the contents of the latter are removed, the empty tins are removed to a safe distance and piled in a heap. When the shooter has completed his task in connection with the well, he disposes of the empty cans summarily. He attaches a fuse to, and fires, the heap. Although the quantity of explosive lingering in each tin is insignificant, the terrific devastating forces of nitro-glycerine are brought home very vividly; the empty tins are blown to atoms, while a fairly respectable hole is torn in the ground where the pile stood." (Talbot 1914)

Once the well settled down after shooting, there would be a quiet flow of oil bubbling from the top of the well. The rate at which a well "came in" was usually figured from the amount of oil that flowed (or was

Nitroglycerine can, Bradford Oil Basin.
Collection of the author.

The Bradford Oil Basin

swabbed or bailed in lesser wells) in the first twenty-four hours. This "flush" production could be counted on for the first few weeks or months. After that time a driller could expect a lesser "settled" production rate.

Local drillers quickly noticed that Bradford wells not only responded really well to shooting, but that their wells could be shot again and again-when production dropped off- with the same good results. There were two reasons for this strange phenomenon. First, with time, the wax in the paraffin-based oil tended to clog the pores in the sand around the drill hole; nitroglycerine got rid of the wax buildup very rapidly.

Second, Bradford oil was trapped in very small pore spaces in the sands, and adhesion to the grains restricted it from migrating. The shock wave from explosives shook the oil loose from that adhesion, and allowed it to flow to the well. Thus, in Bradford, the bigger the explosion, and the more often, the better. By 1881, multiple ninety quart shots were being made. It has been conjectured that between 1875 and 1900, most of the amazing Bradford production numbers were a direct result of this massive trend of overshooting wells.

One of the largest shots on record was a four hundred quart explosion set off in 1903. The well shot was the Colegrove No.29 on the Wheeler farm, north near Bolivar, New York. Twenty thirty-inch shells, each holding twenty quarts of nitroglycerine were lowered into the well, and the shot was tamped with an eight hundred foot column of fluid. The force of the huge explosion flung a whirling black mass of debris to a hundred and fifty feet above the derrick top. The cleaned well was "an excellent producer" for some time after the shot.

At the heart of well shooting was Col. E. A. L. Roberts of New York City, who, in 1866 had been granted the basic patents on well shooting torpedos. Roberts claimed that the concept of shooting had come to him while watching exploding shells during the Civil War.

The Bradford Oil Basin

He experimented early on with gunpowder shells, but quickly settled on the more explosive nitroglycerine in 1867. That Roberts actually conceived of the well torpedo first is doubtful, but he came into the business armed with patents and the blessings of the court. Roberts's torpedos were immediately in huge demand. But because of the prices the Roberts Company charged ($100 to $200 plus a royalty on the increased output of the well) 'and the simplicity of making the shells, small competing companies immediately sprang up to compete with the new monopoly.

In January of 1871, the Supreme Court again supported Roberts, and granted a permanent injunction against all infringers of the Roberts patents. The Roberts Company began to collect evidence against illegal competitors, and proceeded to take them all, one by one, to court. In this long battle, thousands of suits were filed, and the Roberts Company won them all except the last. The Bradford Era during this period carried a daily front page column listing each day's injunctions against local producers. It was not until May 20, 1883 that Roberts's patents expired, and the monopoly was officially ended.

During this time, however, the staunchest competitors stood firm, and there was a major growth of illicit nitroglycerine factories in the region. Illegal shooters took to placing the well shots at night to avoid litigation. These "moonlighters" did their dangerous job well, and actually made many improvements in the shooting technique, including the development of the go-devil. Shooting wells was a delicate task at best, and there are many tales in the oil regions of men and wagons being blown to bits by their volatile cargo. The nitroglycerine was always on the verge of being unstable, and the slightest jolt could set it off at any time.

The liquid was so sensitive to agitation that the eight quart metal cans in which it was transported were capped with corks. There had been too many bad experi-

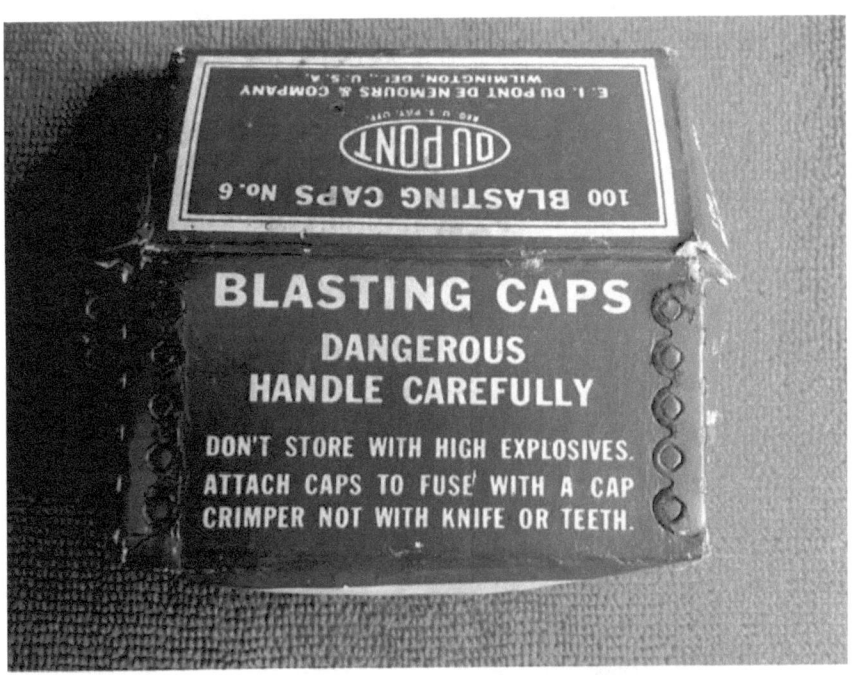

The Bradford Oil Basin

Shooting an Oil Well, Bradford, Pa.

ences where turning a metal screw cap had cost a man his life.2 Moonlighters filled their shells in darkness from metal cans strapped to their backs. It was an incredibly dangerous job followed under the most dangerous conditions possible. It was popular, however, and "the ranks of this illegal profession expanded and thrived for almost a decade" before the feud between Roberts and the producers came to an end.

After 1883, shooting reverted to open enterprise, and many of the old underground companies continued on openly with good success. With time, the shooter's wooden wagon was replaced by an automobile, and improvements in explosives manufacture made the shooter's life a bit safer. Standardized formula nitroglycerine was more stable than the bootleg product of the illegal years, and the shooter was much safer working in daylight. The process of shooting, however, continued basically unchanged around Bradford until after the Second World War, when hydro-fracking replaced nitroglycerine in the Bradford fields.

The Bradford Oil Basin

From the Eldred Eagle, September 12, 1950:

"Eldred, PA - Tragedy stalked into the very heart of Eldred, snuffing out the lives of two local residents and causing heartfelt grief to more families around the vicinity. At 9:35 a.m., Tuesday morning, the dynamite blast ripped through the mix house, completely demolishing it and badly damaging nearby buildings. Four additional employees were injured and treated at the scene as guards kept newsmen and photographers away from the blast area.

Area residents ten miles away heard the booming explosion. Windows were shattered within a two mile radius. Everybody knew almost at once what had happened. The little community of 1,200 people is every conscious of the powder plant danger. Within minutes highways leading to the company property, about one mile north of town, were clogged with traffic. At the time, National Powder was the largest industry in town employing sixty-five persons.

John McCord, of Duke Center, was sitting in the Sandwich Shop in town, sipping a cup of coffee when the blast rocked the town. "The patrons thought the windows would be pushed in and they moved quickly away," he said. "The windows didn't break but they sure did shudder for a few seconds. People began pouring into the main street and within a matter of seconds the cars leading to the plant were bumper to bumper.

The Eldred Fire Department and ambulance responded and remained at the plant throughout the day. Dr. Fred Gabriel administered first aid to the injured. Ambulances from Kane, Port Allegany, Portville and Olean, NY, also responded. Coroner Beatty of Bradford conducted the investigation and released the names of the dead by Tuesday afternoon.

Employees at the Artline Corporation, a furniture manufacturer plant across the valley from the scene, reported windows broken in the plant. "Sawdust was kicked up like a swirling snow storm and people got hysterical. We put our hands over our heads because we thought the walls were going to cave in," said one employee.

Some of the victims were blown to bits as the blast destroyed the steel structure where ingredients are mixed in the production of dynamite. Tearfaced relatives and friends stood silently along the highway and near the company offices as Pennsylvania State Police, Coroner Beatty and undertakers began the grim task of combing the ruins for bodies for identification, which took hours.

It was reported that body parts were found in the trees around the blast area."

The Bradford Oil Basin

Jon C. Cawley

Chapter 5
The Era of Tanks and Storage

Tankage

Once Pennsylvania crude flowed to the surface, it had to be stored and shipped to refineries and markets outside the oil fields. Originally, the small quantities of crude available had been marketed for medicinal use, and as a simple lubricant. But soon, the invention of kerosene lamps-John Austen's "Vienna burner" of 1855- greatly increased the demand for petroleum. After 1860, oil was refined mainly into kerosene for lighting. Slowly but surely, petroleum replaced sperm oil from the New England Coast, and Pennsylvania production sounded the death-knell of the Atlantic whaling industry.

Early oil from the more southern Pennsylvania fields was floated on rafts down the Allegheny river to Pittsburgh, or carried to other markets by teamsters who braved the rough terrain and muddy roads of the time. Early producers stored the oil in plank-lined pits in the ground, in rough wooden boxes or tanks, or in barrels to be transported. It was a young and impromptu industry. Barrels were nonstandard (many were salvaged kegs, or whiskey and pickle barrels) and they leaked badly.

Much oil was also lost from the open storage pits and tanks. The tanks leaked, and being open to the air,

The Bradford Oil Basin

The Bradford Oil Basin

the more volatile parts of the oil quickly evaporated away. Open storage also constituted a major fire hazard, especially since it was usually located within the working oil fields.

By the time the Bradford fields came into production in the 1870s, the technology of transport and storage had improved, although irregularly, across the region. The earliest pipelines had been laid, but were hardly perfected. The first metal storage tanks had made their appearance. And both wooden and iron railroad tank cars were beginning to see use in moving the oil.

The basic techniques existed, but it took the development and production of the Bradford basin to pull it all together, to unify the industry into a modern form. When reporters described the oil regions in 1854 as being on the verge of coming forth in a blaze of glory, they were foretelling events that soon came about in Bradford.

Within five years of the first producing wells in Bradford, production was at 25000 barrels a year, and was steadily increasing. Bradford had the distinction of being the first field in the world to yield 100000 barrels of oil per day. There was an order of magnitude more oil being produced and sold in Bradford than anyone had ever seen before. And it was a great challenge for the new city to handle the flow.

The most logical oil storage for many Bradford producers was the wooden tank. Oil field supply companies by this time were offering good quality wooden tankage. Lumber (in the early years) was inexpensive, and the wooden tanks were easy to ship and assemble in the fields. Wooden tanks were made in all sizes, and came either open-topped or covered. They were built from either clear pine or redwood; the boards were cut with a correct bevel, and with tongue and groove edges to reduce leakage.

Wooden tankage with iron bands.

Early iron tankage.

The tanks were bound with iron hoops and draw lugs, and the finished units were sealed with okum- one of the few sealers of the time that the oil wouldn't dissolve. Details of tank construction varied, but general construction forms were followed. The Tuna Manufacturing Company of Bradford, for example, suggested the following specifications:

> "Wall thickness, two inches for 20000 gallons or less; two and a half inches for over 20000 gallons. Thickness of staves and bottoms to be onefourth inch less than above. Standard inside depth, one foot seven inches, to fifteen feet seven inches. Hoops, five eighths inch round with malleable iron draw lugs; spacing to give a factor of about four to one. Lumber to be thoroughly dry without loose or unsound knots."

But wooden tanks, especially the larger ones, were not without their problems. Wood was still a fire hazard, and the larger tanks over 1200 barrels had a bad tendency to burst. There were definite limits to the strain that wooden tanks would bear. And there was still a problem with evaporation as well; the tanks were usually coated on the outside with a dark, heavy oil derivative to further discourage leakage.

This black exterior absorbed heat, and the warm oil evaporated more quickly from the open (or at least unsealed) top. The larger the tank, the more heat it gathered, and the more volatiles were lost. Wooden tankage, practical in smaller sizes, became questionable as its size increased. Still, wooden storage was generally practical for derrick and lease work, and smaller wooden tanks have remained in use on pumping leases until the last few years. Numbers of them can still be seen on abandoned leases in the region.

As the oil came out of the ground, more and more tankage was needed. Knotless lumber became scarce quickly, and since larger tanks were in demand, iron became a possible answer. Soon, great numbers of iron tanks were being built near the central shipping points

The Bradford Oil Basin

of Bradford and Olean. Outside financiers, attracted by the flow of oil and money, had already noticed the Bradford region. And in an early move to control a lion's share of the oil storage and shipping, the Standard Oil Company (under John Rockefeller) built much of the new style tankage in the Bradford/Olean area.

Iron tanks were built as round cylinders of caulked and riveted iron plates. These plates, five feet by ten feet, were riveted into rings to a height of about thirty feet. The plate thickness tapered from a half an inch in the bottom rings, to a quarter of an inch at the top. Because of the strength of the iron, the tanks could be built to exceed the largest of the wooden ones.

Bradford basin tanks usually held about 35000 barrels each. Some of the early iron tanks were made virtually air-tight, but since they quickly gathered heat and built up vapor pressure, they tended to pop their rivets and leak badly. Breather valves were quickly added. Other innovations included improved filler and discharge pipes, draw-off valves at the bottom to remove water and sediment, and hatches on top for entry and ventilation when inside work was necessary. For a long time the iron tanks were painted red which, although better than black, still encouraged heating. It was some time before oilmen began to paint their oil tanks with white enamel to discourage heating and keep their oil cool.

The iron tanks were built together in groups, and the Bradford countryside was the original home to the "tank farm" concept. The local political scene and company interests had some input into the development of the tank farms; it was natural to keep one's oil holdings together in one place on one's own land. But there was another side to the issue as well. To be stable, the larger tanks had to be built on flat and solid ground.

The 35000 barrel tanks weighed some 1260000 pounds and sagging or sloping land could mean the loss

Burning Oil Tank, struck by lightning, near Rock City, N. Y.

The Bradford Oil Basin

of a tank and a good amount of oil. In Bradford, much of the solid ground was along the Tunungwant Creek, or on the flat hilltops. In New York State, much of it was in the wide, gravel-filled river valley of the Allegheny. The tanks were built on sand and concrete foundations, and were surrounded by a four foot earthen dike to catch most to the oil if the tank should burst or boil over in a tank fire.

When Bradford became the home of hilltop tank farms and iron tanks, it also gained a reputation for a peculiar sort of local oil disaster. Although the iron tanks lacked the flammability of the older wooden tanks, they had electrical properties which were little anticipated early on. Sitting exposed on the tops of hills, they attracted lightning during storms, and set themselves afire. Tanks in the valleys also caught fire from time to time-the valleys were home to the railroads as well, and sparks from a steam engine could quickly set one or more tanks ablaze.

Tank fires were always of great interest, even during the years that they were relatively common. As the oil burned from the top, the iron walls of the tank conducted heat to the rest of the oil. If left alone, the burning crude would finally boil-this made the burning more violent, and the tank would sometimes burst or explode. When an oil fire struck, spectators would come from great distances, guided by the tall black pillar of smoke and flames that reached up from the tank.

Naturally there was a set procedure for fighting such fires. First, firefighters would begin to pump oil out of the tank, and water in. Pumping continued until the bottom rings were covered with water, or until the heat became too intense. Then a cannon was brought in, and cannonballs were shot into the middle regions of the tank in hopes of draining off the rest of the oil. In this way, most of the tank could be salvaged for repair. Often, as many as six cannonballs were shot into a large tank.

TANKS THAT WERE STRUCK BY LIGHTNING AT EAST KANE, PENNA. ACH TANK HOLDS 35,000 BARRELS OF PETROLEUM

The Bradford Oil Basin

Later, fire-smothering foams proved more effective than cannonballs, and foam systems were installed on most of the larger tanks. These foams were mixtures of various coagulents, bicarbonate of soda, sulfuric acid, and water. The wet, carbon dioxide-rich foams would cut the burning surface off from the air, and smother the flames quickly.

As electrical grounding techniques have improved, and as the railroads eventually withdrew or changed over from steam to diesel, tank fires now seldom occur around Bradford. Many of the riveted iron tanks in the region remained in use for over fifty years.

In 1892, the word "steel" was first used in connection with oil tankage. The mention came, as might be expected, from The Oil Well Supply Company, a subsidiary company of U.S. Steel in Pittsburgh. Welded steel tanks came into use around Bradford some years later. The advantages of a cheap, welded, one-piece tank were obvious, and most of the tanks used now are made of welded steel.

The Barrel

On a smaller scale, the barrel has always remained as the basic unit of oil storage and transport. In the early years, barrels were scarce; it took time for professional coopers to realize how lucrative their trade might be in the oil regions. Even then the quality of the barrels remained questionable. To turn out mass quantities quickly, the coopers often used cold, green wood that was not steamed, washed, or dried. The decent barrels went more often to refined oil products, and the rest went to the producers for crude (Once a barrel was used for crude, it couldn't be used for anything else.) The cost of barrels fluctuated, but was never really low-$3.00 each was an average price. Thus, it was not uncommon for barrels to be worth more than the oil they contained.

The Bradford Oil Basin

The barrel was standardized at forty-two gallons in 1866, just a year after Job Moses drilled his first well at Allegany. A meeting of producers in Venango County in August of that year saw standardization as a necessary step for trade and shipment. The odd forty-two gallon size was unanimously voted in; forty gallons, the standard unit, and two gallons for leakage and evaporation. Thus, from a humble start, salvaged from the basements of stores and taverns, the barrel rose to become the world standard for oil measurement.

As large storage, railroad transport, and pipelines came into use in the Bradford years, The "barrel" became as figurative as it was tangible. The oil exchange, built in Bradford in 1878, was the most active in the world, by 1881 was doing a million dollars of business a day in bulk transactions. Tank volumes and flow rates were calculated arithmetically, and much Bradford oil never actually saw barrels while it was in the oil regions. Still however, even today, many producers still pack their smaller outputs of crude into barrels to go to the refineries.

Steel containers began to replace the traditional coopered wooden barrel at about the turn of the Century. The first steel barrels were imports from Europe, usually containing chemicals. But as the local refineries began to use them they were soon produced by American companies as well. The first of these appeared about 1902, and they were common by World War One. The steel barrel later evolved into the steel drum, with its straight sides more convenient for shipping.

Railroad Tank Cars

Because of the price, and the limits of the old barrels, it is not surprising that shippers turned to bulk methods of moving oil. The railroads took an early turn toward bulk by introducing oil tank cars.

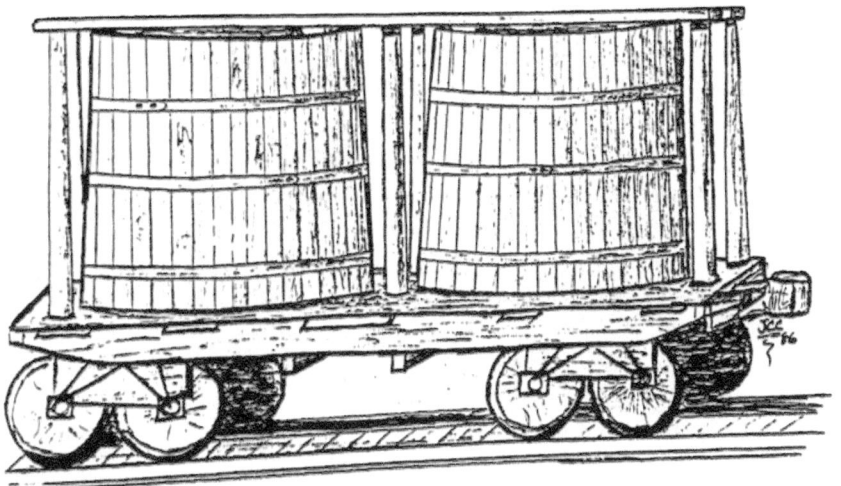

Early Densmore oil tank car.

The Bradford Oil Basin

In the summer of 1865, Amos Densmore of the Oil Creek area mounted two, forty-five barrel wooden tanks on a regular railroad tank car. This first bulk oil shipment was sent on the Atlantic & Great Western Railroad; the car passed through Salamanca, New York, on its way to New York City. The experiment worked, and railroad tank cars quickly took their share of the oil shipping trade. Other railroads adopted the Densmore tanks, and by the spring of 1866, there were hundreds in use. They were kept in regular traffic on the lines until after 1871.

Like most wooden tanks, the Densmore style cars had a problem with leakage, and they didn't hold up well in train wrecks. The upright tanks often burst when jolted, and, with their high centers of gravity, the cars tended to tip over in an accident, adding flaming petroleum to an already bad situation. Iron tanks were tried, to reduce leakage and bursting. But the heavy iron only added to the high center of gravity problem, and the iron Densmore cars were more top-heavy than the wooden ones.

The first successful horizontal iron tank cars came into the oil fields around 1868; still before the beginning of the main Bradford Boom. The new tanks consisted of a horizontal "boiler" mounted solidly on a railroad car frame. The new cars held between eighty and ninety, and later, one hundred barrels of oil. Although more expensive, the boiler type cars were much safer in transit. The iron cars had an additional innovation that was quite important.

The tanks were fitted with a dome on top that allowed for the expansion and contraction of the oil without damage to the tank. The railroads recognized early the importance of safe and solid tank cars- the horizontal cars had a low center of gravity, and were impervious to most shocks and jolts. Leakage was minimal, and the expansion domes saved oil and increased the expected life of the cars.

Jon C. Cawley

By 1872, the new style tanks were generally the rule on the oil railroads, and the older cars were on their way out. Most sources report that the last of the wooden cars had disappeared by 1880.

Later tank cars were made of steel, but the styling of the cars has changed very little since they appeared. The long cylindrical cars are still an everyday sight on the railroad lines in the Bradford region.

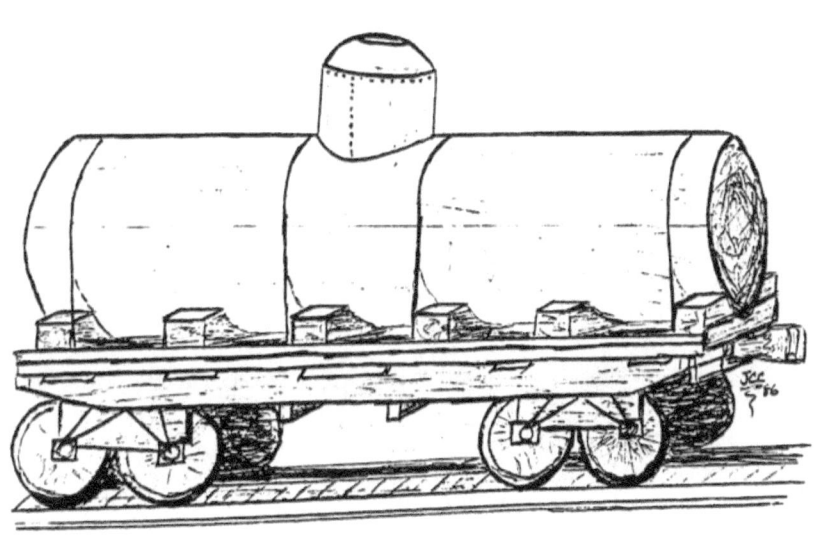

Early horizontal iron rail tank car.

The Bradford Oil Basin

Jon C. Cawley

Chapter 6
The Era of the Pipelines

Pipage

Oil pipelines began as an experimental alternative to exorbitant prices charged by teamsters in the southern fields. They ended up in the Age of Bradford as a major political bugbear between the producers, the railroads, and the Standard Oil Company.

The first Pennsylvania oil pipelines were thousand foot long feeder lines from tankage to nearby refineries. They were built of cast iron piping, and were driven by the force of gravity. The primitive pipe of the 1860s had major limitations. The pipes leaked, and suffered badly from seasonal expansion and contraction. Many of the lines never saw seasonal problems- the teamsters destroyed or damaged the pipelines when they could. By 1863, however, there were at least three short pipelines functioning in the lower fields.

By 1874, the idea of longer lines was in vogue, and the Columbia Conduit Company---owned by David Hostettor, a patent medicine maker and oil entrepreneur- as attempting a three-inch in diameter pipeline from the southmost oil regions to Pittsburgh. Within the thirty-five mile stretch, Hostettor's men met quick opposition from the B&O Railroad, who naturally refused the competing pipeline to pass through any of its rights-of-way.

The Bradford Oil Basin

The men would lay the pipe, and the B&O would tear it up as soon as it was laid. The matter was not settled until the next year, when D. McKelvy & Company leased the pipeline from Hostettor. They solved the problem by taking the crude oil out of the pipe, carrying it by wagon across the railroad land, and putting it back into the next section of pipeline.

The first pipeline in the Northern fields was built by the Empire Transportation Company of Philadelphia in 1875. Empire was a subsidiary of the Pennsylvania Railroad, who was now beginning to compete against the Erie Railroad for oil shipments from the Bradford basin. The pipeline was fourteen miles long, and ran from Tuna Valley to Olean. Its path carried it over Rock City at an elevation of 2350 feet; there was a large pump station at the Tuna Valley end, and a booster station at the bottom of the hill at Rock City that pushed the oil up and over to Olean. The building of the Olean Pipeline was considered a major engineering feat—and the pipeline itself was built in eighty days. The line was tested to 1200 pounds of pressure, and was constructed of two inch-in-diameter wrought iron pipe with screw couplings.

Meanwhile, by the 1870s, John Rockefeller had turned his attention toward the Pennsylvania oil fields. Rockefeller saw the chaotic way that the oil business was being run, and decided that it would be profitable to Standardize the oil industry. Oilmen, he reasoned, should not work under competition with each other, but rather by "cooperation"- preferably under his own direction. The best way to control the industry was by seizing it at the bottleneck- in transportation-and refining. Refining he handled under the Standard Oil Company; pipelines through the United Pipeline Company; and everything else through the imfamous South Improvement Company, which was to deal mostly in railroads and freight.

Jon C. Cawley

In 1871 the South Improvement Company united the railroads in a scheme to divide up the oil traffic among themselves, and under its own direction. Pennsylvania producers found out about the new exorbitant shipping rates prematurely, and at a 2000-producer meeting at Titusville, voted an immediate oil embargo. The railroads disowned the whole scheme, and the Pennsylvania legislature revoked South Improvement Company's charter.

The company vanished in a puff of smoke. Rockefeller and his men, meanwhile, simply regrouped and carried on with their plans. Standard and United continued to buy refineries and pipelines, and to make deals with the railroads. By 1875, the Standard had growing influence with Erie, the Atlantic & Great Western, and the New York Central. Bit by bit, the fields fell more and more under Standard control.

In November of 1877, the second Bradford basin pipeline was built to help carry the increasing flood of oil. The Carrolton line was twelve miles long, and ran from Tarport and Tuna Valley to Carrolton, New York, where the oil was held for shipment in great iron storage tanks. The pipeline was built of four-inch wrought iron pipe. The pumping station was at Tarport, and the line had a capacity of 8000 barrels a day.

The line hired the hydraulic engineer Henry R. Worthington to work out problems in the pipeline and pumping station. Worthington applied multi-cylinder pumping engines which worked more smoothly than the old single stroke pumps. His work in the Bradford basin greatly revolutionized oil pipelines in general.

But times were getting tough politically, and the squeeze of the Standard was becoming tighter and tighter. In 1877 the Standard people incorporated the Olean Petroleum Company Ltd., whose major interest was the Olean pipeline.

The Bradford Oil Basin

The Empire company gave in after a fight, and the Olean pipeline was under United control by the end of the year. By 1878, United owned the Carrolton pipeline as well. This move secured a Standard monopoly in the Bradford region.

In 1878, a ten mile long, four inch extension line was added to the Carrolton pipe. This extra access ran to Salamanca, New York; two 25000 barrel storage tanks were built there, and the capacity of the line was increased by thousands of barrels per day. United pipeline was omnipotent. "Wise men bought Standard stock and hung on.

For awhile the cheap transport prices appeased many Bradford producers. Standard, with its extra storage, allowed for banking by oil certificate- producers could take their oil to Standard tanks, store it free, and play the fluctuations of price. They would sell at slightly higher prices as the oil was needed for shipment by Standard.

Some of the more wary producers had talked early about an independent pipeline from Bradford to Buffalo, and in 1877 a possible route was surveyed. The independents thought to break the Standard's monopoly by hooking into the aging Erie Canal system. But Standard quickly bought out the main interest in the pipeline project.

The plan was again organized by independents in 1878 under Bradford's Lewis Emery Jr. (one of the region's most vocal opponents of the Standard.) Their Equitable Petroleum Company built the pipeline. The Erie Canal, however, was seasonal in operation. The seasonal alternative for getting the oil to the East Coast from Buffalo was shipping by rail; and most of the railroads were still under Standard control. Producers began to look to the South and East:

The Bradford Oil Basin

"It requires no great power of foresight to comprehend the construction of seaboard pipes through which our product shall flow to the east and south. The time is coming when this product will be prepared and shipped as distillate and railroads might as well come to the conclusion one time as another." (Bradford Era, January 1878.)

By this time, more producers were becoming dissatisfied with the Standard. As production rose higher each month, the Standard began having problems handling all of the oil. In forming a monopoly, it was taking a responsibility upon itself to keep the oil flowing from the fields to the markets. But the Bradford fields were producing at a more than prodigious rate. Soon, the Standard realized, it would have to expand the size and number of its pipes reaching from the Bradford fields. But first, it had a ready chance and excuse to raise its profits, and finance the later piping.

As the monopoly was becoming set, Standard announced that it would take no more oil except that which could be shipped immediately. Producers, of course, did not take well to this, because it disallowed playing the storage and market advantage that they were used to. It also suddenly put the problem of storage back into the hands of the producer. A Bradford Era reporter in a notice about Gilmor City stated in January, 1878:

"Never since I have been in Oildom have I seen tanks in such great demand. Oil flows from the tanks, and is being gathered by the dippers only to be again thrown upon the waters to float away. The secret of this is the prices of tanks have advanced 25% in the market values within a couple of weeks. 200 and 300 barrel tanks are springing up all over the field as if by magic. It is hoped that this state of things will not long exist, and that better times will soon come to our relief. (Bradford Era, January 1878.)

The Standard, however, pressed even further. In January, J.A. Bostwick Company, the Standard buyer at Bradford announced that they would take no oil whatsoever except at prices lower than the market quotation.

The Bradford Oil Basin

Adding insult to injury, they insisted that producers stand in line personally to sell their oil at the reduced rates. The independent producers of Bradford would only stand for so much, and almost immediately letters openly against the Standard were being published in the Era and other newspapers:

> To the editor of the daily Era: Sir-I am led through the courteous manner m which you treated a former communication to presume further upon your patience. A few weeks ago the producing public was informed that tankage was everywhere full, and that it was impossible for the Standard Company to remove all the oil produced in the district. Presently we, are informed that in order to have our tanks relieved we must dispose first of their contents and at a double pipage "off." Strange to say since that time it has not been found necessary to have a tank in the district overflow. The Standard Company immediately on adopting the discriminating process discovered (?) that there still remained sufficient room to receive all the oil that the district could produce. From thesecircumstances is it not safe and proper to assume that the old system of discriminations against this district has been inaugurated, but under a different process? Can it be possible that the producers hereabouts are all blind to these facts, or a few of the most prominent "purchased" for the time being for purpose of easing down the burden? May we not look in the future for continued discriminations against this field? And was not the adoption of the "United" plan for a little season a scheme to invite the innocent producers to "peace" at a time when war would have been dangerous? Are producers aware that the Standard is now engaged trying to close up the last chance of a pipe to Buffalo, as well as procure legislation against the opening of the Genessee Valley Canal? These questions, Mr. Editor, I leave for you or anyone else to answer, should they be considered of sufficient value. Young Producer (Bradford Era, January 31,1878)

The time for an independent pipeline to outside markets had come.

The Tidewater Pipe Company was formally organized in November of 1878. The founders of the pipeline company included Byron D. Benson H.L. Taylor, A.A. Sumner, and David McKelvy (of Columbia Conduit fame). The original plan called for a great six-inch pipeline from the Bradford fields to the New Jersey seacoast. Negotiations with Franklin Gowen, president of Reading Railroad, however, insured tank car transport from

Williamsport, Pa. to the coast. And so the plans were changed in favor of a shorter line to Williamsport, still 109 miles long.

Reading pledged money and support mainly because it was involved with its own feud with the Pennsylvania Railroad over coal. The only trade-off made by Tidewater was a pledge that it would build no further than the Reading loading site at Williamsport for eight years.

Such a major pipeline had never before been attempted, and skeptics called the project "Benson's Folly." The iron pipe sections were made by the Reading Iron Works and National Tube Company and were seventeen and a half feet long, longer than anything used before. The line was laid on private rights-of-way across the Allegheny mountains (2600 feet of elevation in between) in the winter of 1878-9. And despite bad weather, technical problems, and threats by the Standard, the line was finished by its original deadline in May.

Benson designed new 'triplex' pumps for the pumping stations at Corryville and Orbisonia. These pumping plants were larger than any used before, and were run by multiple steam engines. By the time the line was completed on May 22, 1879, nearly half a million barrels of oil were stored at Corryville, while 60000 barrels of storage and a long string of Reading tank cars waited at the Williamsport end. The oil flowed smoothly, and reached Williamsport on June fourth. A three foot log and a coil of rope were removed from the pipe, and the Tidewater pipeline became operational.

With the opening of the Tidewater pipeline, freight rates in Bradford fell overnight from eighty-five cents to thirty-five cents per barrel. The Tidewater could get oil to the coast for about thirty cents a barrel. Standard quickly invoked another "shipments only" policy to pressure the producers. Convinced of the necessity of long distance pipelines, however, it started several long line pipes of its own.

The Bradford Oil Basin

The Standard Oil Monopoly.
From an early post card.

The Bradford Oil Basin

The Standard began construction of an eighty-mile conduit south from Bradford to Jersey Shore, Pa. along the tracks of the Pennsylvania Railroad (This was extended to Philadelphia later, under a National Transit Company title). And another was begun toward Buffalo on the Buffalo and Southeastern line. These pipelines were in operation by the end of 1880.

Then under the name National Transit Company, Standard set out to build a pipeline the entire 400 mile distance to the New Jersey seaboard. With some dealing, Standard got permission to use the right-of-way of its old friend, the Erie Railroad. This reduced private land problems, and insured fewer problems with topography than Tidewater had faced. And being close to the Erie route, the oil could be taken from the most recently completed section of pipeline and transferred to Erie rail cars at any point along the way.

Construction began in the fall of 1880, from Olean eastward. By June of 1881, the line had reached about 100 miles to Painted Post, New York, and construction was being started from the other end- at the Standard refineries of Bayonne, New Jersey. The full length of the National Transit pipeline was kept in New York State, where a free pipeline bill, passed in May of 1878, now made it legal for the Standard line to cross railroads, canals, highways, and farms. The line was kept to the North, outside of Pennsylvania, where the free pipeline bills had been suppressed.

The Standard line consisted of two, six-inch pipes, and was tested to 1500 pounds. It was constructed of lap-welded wrought iron with heavy collars. The ends and sockets were threaded nine threads to the inch to prevent leakage. The pipes were buried to a depth of eighteen inches for most of the line's length. And there were eleven pumping stations on the final, 400 mile pipeline. These stations were each equipped with double sets of boilers, engines, and pumps in case one set broke down.

Jon C. Cawley

The pipeline was finished to the coast in 1881, claiming for itself the distinction of being the first long distance coastal line (Tidewater reached the coast in 1885.) National Transit's main line pressure stood at 900 pounds, and the line pumped 40000 barrels of oil a day.

By the time the two major pipelines were in use, there were a sizable number of independent short lines within the Bradford Basin itself. Most of these smaller lines were gathering systems from the scattered oil towns, and from wells up in the smaller valleys. These lines too were built of cast iron pipe, and most worked by gravity flow. Steel pipe was not used in the pipelines until the 1890s, and the first time that line pipe was screwed together by machine rather than by hand was in 1912. Oil from the gathering networks was shipped out of the regions, or was refined at the local refineries that were beginning to spring up.

In 1892, Lewis Emery Jr. organized the United States Pipeline Company to build a line to ship refined oil products from the local independent refineries. This pipeline would carry "distillate" kerosene to the seacoast enabling local refineries to compete with the Standard. The company was incorporated in Pennsylvania on September 20, 1892. The United States pipeline was to consist of two, four-inch pipes, one for products and the other for crude oil. The line was to run to Hancock, New York, about one hundred miles from New York City.

As the line approached Hancock in December, 1892, it was stopped by armed guards from the Erie Railroad. Emery had the New York State free pipeline bill behind him, but rather than waste time in court, he backtracked to Athens, Pa., and ran his pipeline south to Wilkes-Barre, Pa. instead. The lines reached Wilkes-Barre in the spring of 1893. The new line could ship about 2000 barrels a day. The line was later extended to Philadelphia, under the Pure Oil Company name, and still as an ally of the independent refiners. Although the pipeline

The Bradford Oil Basin

had only a slight effect on the overall oil industry, it did actually stand as an independent entity against the Standard Company.

After the peak of production in 1881, the great pipelines began to fall on hard times. As less oil was being shipped from Bradford, the lines had problems keeping capacity filled. Standard kept its lion's share of the Bradford output, and Tidewater, who was still building toward the coast was left to struggle. Tidewater reached the coast at near financial ruin in 1885, two years after submitting to a market compromise with Standard. The market-sharing compromise in 1883 called for eighty percent of oil shipments to Standard, twenty percent for Tidewater.

Standard had come out on top, but the major part of the game was already over. The National transit line carried shipments of oil for many years, until October of 1927, when Standard closed the line and moved on to richer fields. Tidewater has shipped oil in at least portions of its lines until the present day. Other pipelines have been built and used since, but the age of the Great Bradford Basin pipelines was over by 1929.

The Bradford Oil Basin

Jon C. Cawley

Chapter 7
The Era of Pumps and Pumping

Pumping Wells in Bradford

In 1881, the flush production of the Bradford region peaked. Bradford was a prolific producer; for every twenty wells drilled in the district, nineteen of them had struck oil (compared to about four in twenty striking oil in other fields). And many of these were flowing wells, which gave oil under their own power. But with enough holes into the earth, the gas pressure in the oil sands began to decrease, and the wells stopped flowing.

These were added to the sizable number of wells that had never flowed, and pumping became an important part of the business of Bradford. The last flowing wells (with the exception of the Music Mountain Pool 37 years later) were gone by the 1890s. After 1881, hundreds of new wells were still drilled each year, and old wells were cleaned out and pumped, but the boom had leveled off, and declines in production quickly followed.

After wells were drilled into the producing oil horizon, they were soon tubed----the hole was sealed against water, and made ready for pumping. This was done by lowering two inch-in-diameter copper tubing to the bottom of the hole, and sealing it between the upper wet horizons and the oil sand with a packer or flax-seed bag. Usually the bottom end of the tubing carried an oil strainer to cut down on sand and debris settling into

The Bradford Oil Basin

the well pipe with the oil. Often the tubing and packers had to be pulled and readjusted several times before the well could successfully be pumped.

Once the well was sealed, a rod arrangement with a plunger or pump on the end was lowered into the tubing. This vacuum pump would lift oil up through the tubing to the surface as the rods were moved up and down from above.

The earliest of these "sucker rods" were made of long, octagonal pieces of ash or chestnut wood connected together with cast iron couplings. These wooden rods were light, and weighed even less within the liquid inside the well. This was an advantage, since the entire set of rods were lifted up and down from the surface. Wood, however, had a disadvantage in strength. A sucker rod system, like a chain, is only as strong as its weakest section, and the wooden rods were weak wherever they attached to the iron coupling ends.

Iron sucker rods were stronger, but heavier. They were made of hand-forged metal, and had mated screw couplings on each end. Iron sucker rods were standard for a very long time in the Bradford region, and are still in use in many of the older wells. The biggest disadvantage of the iron rods was that they were brittle. Under the constant strain of pumping, they fatigued and broke. The couplings were also susceptible to wear--- they were slowly abraded by sand in the oil, and the rods had to be replaced eventually.

Solid, weldless iron and steel rods were gradually introduced. These mainly reflected technological improvements in metal-making in the 20th Century. The newer rods were made to be uniform and to take the pumping stresses better. Double male ends with separate couplings between each rod section have been introduced. As the wide shoulders wear out, the couplings rather than the entire sucker rod can be replaced.

Early gear-driven pump-jack.
Bradford field.

The Bradford Oil Basin

Jon C. Cawley

The pump plunger rode at the bottom of the sucker rod string inside a smooth pumping barrel in the tubing. The usual plunger set-up consisted of moving and stationary valves. The upper was a moving ball valve which opened on the downstroke and closed on the upstroke. This lifted oil up into the tubing. The lower valves were stationary, and closed on the downstroke to prevent the oil from being pushed back down into the sand. These lower valves were originally made of leather or canvas flaps. As might be expected, plungers have become more complex with time---with various oil savers, special rubber seals, and valve seats.

At the top of the well was the casing head, and another set of oil tubing and seals. At the casing head was an adapter that held the top of the tubing inside the well casing. Above this was a 'tee' which diverted the oil to tankage. Through the center ran a polished brass rod which connected the top of the sucker rods to the pumping machine. And around the polished rod above the tee was a stuffing box. This box was filled with soft packing, which sealed the top of the tubing but enabled the polished rod to move up and down. These parts of the pumping system have changed little or not at all since they were first introduced.

In the Pennsylvania fields, the underground portion of the pumping apparatus had to be pulled quite often. Under the decreased pressure of pumping, the moving oil deposited paraffin on the strainer system, plunger, and rods. If the wax was allowed to build up, the output from the well decreased and eventually stopped altogether. Usually, after some months of pumping, the tools were taken from the hole, and the wax was cleaned from them. They were checked for worn parts, and put back into the hole.

This pulling was done with the derrick or portable drilling machine, with a mast and a team of horses, or with a tractor or special pulling machine.

Well pulling still goes on regularly in the Bradford oil fields, and is considered a normal part of well

The Bradford Oil Basin

maintenance. Because well pulling is somewhat expensive (about $1000 in 1985) this "healing over" of the wells has brought an end to some of the smaller producing leases.

The earliest pumping was done using the walking beam of the well's oil derrick; the sucker rod string was attached to the end of the beam. After a few years, smaller walking beam engines were brought into the wells so that the larger derrick engines could be used elsewhere for drilling. These machines were the predecessors of the more modern pumping jacks. They were powerful-looking machines with complex, exposed gearing and were usually run by steam. By the 1890s, the hills around Bradford were loud with the wheezy sound of oil pumped by steam.

Some of the pumps made louder sounds than others; owners with more than one well often put whistles or "barkers" on the steam exhaust of their pumping engines. This gave each well a different and discernible voice that could be heard from a distance in the woods. If a well stopped pumping, the sound would stop also, and the producer would know to restart the pumping engine.

Central Power Leases

In the more congested fields, it made more sense to try to centralize the power for pumping. The first successful attempt on record had been by E.O. Yates, who in 1880 patented a system of pull rods and rod holddown devices. Yates's system was the basis of most of the later centralized leases.

To centralize pumping, an owner would mount a great engine in the center of his lease. This engine turned a large, horizontal eccentric wheel. Long, iron jerker rods were attached to the central wheel and extended outward toward the surrounding wells. As the wheel turned, the rods were pushed and pulled back and forth; this motion was transferred to each sucker rod system through a pumping jack on each well.

Central Power:
Band Wheel and
Eccentric Wheel
at
Indian Creek.

The Bradford Oil Basin

The Bradford Oil Basin

Pumping Equipment

Bradford Oil Fields

The engines used on leases were major pieces of technology and art. The thirty-five or forty horsepower engines rested on cement pads in the woods and fields, and were housed in corrugated metal buildings. The machines usually used natural gas from the wells themselves (or fuel oil from local refineries) for fuel. They were isolated and independent. They were kept at an even pace by great oily flywheels and polished brass governors. The belts and gears, the band wheel, and the engine made distinctive noises of their own as they pulled as many as eighteen or twenty wells each.

Between the engine and the spider web of jerker rods was the eccentric wheel. These wheels were of two different types. In one version, the engine worked the great wheel directly through gearing and an axle system. The hub of the wheel rested on a large eccentric gear which gave motion to the rods attached around the outside of the wheel. In the other type, the engine moved the wheel by means of a massive belt or band that extended around the whole of the outside rim. The rods were connected to a slightly smaller eccentric wheel attached above or below the main band wheel.

The jerker rods could be attached or detached from the great wheel as different combinations of wells were to be pumped. But it was always necessary to keep the web of lines balanced around the edge of the wheel. An unbalanced power or band wheel did not pump smoothly; it stressed the engine, and it wore the gearwork quickly. The first powers used long wooden jerker rods coupled with cast iron connectors. These were quickly superceded by iron jerker rods. These were one-half to three-quarter inch-in-diameter hand-forged solid rods.

Threaded turnscrews were then included in the rod lines to adjust for length. The most important part of rod transmission was keeping friction down and keeping the rods off the ground. To do this, a battery of hold-up devices and other equipment was developed.

The Bradford Oil Basin

The most simple hold-up was a wooden post with a lubricated slot through which the iron rod was run. These were placed at forty or fifty foot intervals. Better was the tripod and pendulum type of hold-up. The pendulum end rode with the moving rod, greatly cutting down on friction. Holdovers pushed the rods to one side, prevented lines from rubbing against objects and took up slack in the line. Ninety degree link swings allowed the rods to go around corners. And pendulum multipliers conditioned the power to what an individual well needed.

Various sorts of knock-off posts allowed wells to be disconnected from power other than at the central wheel. Rod transmission was reasonably effective and could deliver power to wells up to half a mile away.

The most common central power pumpjacks were the Titan (Oklahoma) types. These machines used an abbreviated walking beam to translate the movement of the rods into vertical lift within the well. The pumping jacks were smaller, about six feet high, and were made of iron. The greatest difference between the types of Oklahoma jacks was the direction the pumpjack faced with respect to the power.

In the Titan, the jerker rods came from the rear of the pump. The lines moved a triangular pivot which lifted the head of the walking beam by means of a pitman. In others, such as the older "standard" jacks, the pump faced the lines. The rods moved a different triangular pivot, which included the walking beam as a part of itself.

An even smaller, straight line pumping jack was sold by Bovaird Company. This device did away with the walking beam altogether. Instead, it used the triangular pivot and a direct vertical lift on the polished rod and well string. This machine was advertised for its "absolutely straight lift" and as "the easiest of all jacks to set." It was never quite as popular as the Titan or Oklahoma jacks.

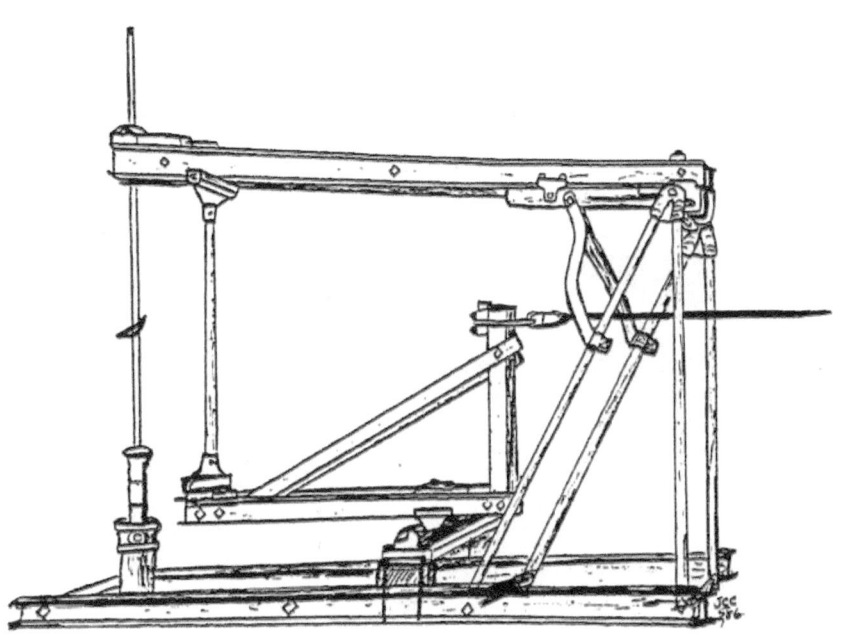

Titan pump jack.

The Bradford Oil Basin

There was one major alternative to rod power transmission for pumping. The other centralized lifting system used steam or compressed air pressure piped to the well head. This "pumping head way" was also introduced by the Bovaird Supply Company. As Bovaird described its operation:

> They are mounted on the casing head or tubing, and require no guys or other support or fastening of any kind. Being of moderate weight and exceeding simplicity, any one having had but the slightest experience in pumping oil can connect them to the well. Let us assume that compressed air is to be used: As soon as a pressure of from 40 to 70 lbs has been generated (this depending on the depth of the well) the valve or stop in the power plant connected to the line leading to the head is opened, and in a few moments the well is being pumped. It is quite immaterial how far the well is distant from the power station... A number of wells can be pumped at one time, this depending entirely upon the size of the compressor used. It is, therefore, plainly demonstrated what a great saving in time can be had. All timbers, or other foundations, bells or other appurtenances are entirely eliminated, thus effecting a great saving in expense. (Bovaird advertising flyer c. 1925)

The major disadvantage to the system was in keeping enough pressure to run the pumps. A large compressor or steam plant was necessary, especially to pump a number of wells. Pressure lines leaked, and steam cooled over distances in the pipes. The system was not greatly popular around Bradford. It was, however, used later by North Penn and other gas companies to pump water from some of their larger gas wells.

Single Pumps
While centralized power was proving a great help on large leases, single pumping jacks were also being improved. Natural gas engines were quickly applied to single pumpjacks). And as diesel and gasoline engines came along by the 1920s and 1930s, they were used as well.

It had been discovered that wells produced better if they were pumped periodically rather than continuously. So many small producers were frugal, and used their newly acquired farm tractors to drive their pumps. In the fields, the walking beam remained as the usual form for pumping. From the great wooden derrick beams to the later smaller metal ones, the system continued to work. And so it was never really changed. As metal gearing for pumpjacks improved, the gear boxes became more compact, and eventually came enclosed. This changed the look of the pumping machine considerably, but not the functionality.

As motors improved, and gasoline and diesel engines were introduced, balance became a prime consideration. The pitman arms were moved forward on the beam, and a counter-weight was added at the rear. This weight compensated somewhat for the weight of the sucker rods in the hole. Weights were also added to the cranks to even out the motion of the pump. And because vertical lift was important, a rounded horsehead was added to the front of the beam. With the addition of a flexible bridle, the polished rod and string would lift vertically, regardless of the angle of the beam.

By the 1940s, electric motors were being used to run single pumping jacks. Electricity had many advantages in pumping. Access to electricity around Bradford was good and was inexpensive. Electric motors needed no refueling, and little maintenance. And with electric timers, the pumps literally ran themselves. In 1940 alone, producers bought some 577 of the new electric pumps from local supply companies. This step marked a trend toward individual well operation that was to continue to the present.

Centralized electric powers were later introduced; the decline of natural gas availability was becoming a problem to producers with the large gas engines. After the second great production peak in 1937, however, the large central leases were beginning to be altogether less economic.

Electric pumping jack.

The Bradford Oil Basin

Jon C. Cawley

With water flooding, the immobility of the central powers was a disadvantage. Where gas and oil remained available, central pumping continued. Some of the leases were actually converted to electricity---but very few.

As flooding and individual pumps became more popular, continuing after World War II central powers slowly approached extinction. There were decreasing numbers of them through the 1950s and 1960s. Currently, Pennzoil maintains a few relic central power set-ups that produce oil. And there are some independents who work old central powers on their land occasionally. Most of the large central powers, however, are rusting in the woods, abandoned.

Electric pumping was made to order for small producers, and for flooding operations. And today the electric pumpjacks are a familiar landmark in the Bradford area. They are scattered in fields among the grazing cows. They can be found in side yards and in the parking lots of buildings in Bradford City. The pumps nod slowly up and down with a thudding sound. And on winter nights they make a soft electrical whining sound in the cold.

The era of pumping continues in the Bradford district.

KENDALL REFINING COMPANY, BRADFORD, PA.

The Bradford Oil Basin

Main St., Bradford, Pa.

Glass Whale-oil fount.

Jon C. Cawley

Chapter 8
The Era of Refining and Secondary Recovery

Initially, crude petroleum as a natural resource was fairly straightforward--aside from native uses as a medicine and preferential natural laxitive, heavy body crude oil was handy to provide lubrication for the mechanisms and moving parts of naval machines, large levers and construction mechanisms, and the steam engines of the western industrial revolution. Oil-soaked rags had quickly become useful as lubricant and as packing for cylinders and pivot points. Finer grades of oil were recognized as being useful as lubricant for smaller mechanismn, clockworks and gear trains.

By the mid-century of the 1800s, Railroads and industrial steam engines were requiring more and more friction-free surfaces and at higher and higher temperatures. Oil worked well with hot steam engines, since for the most part, oil and water did not mix.

Natural Gas

At the well-head, oil was unusual in that, when the natural gas pressure (itself flammable) was released, the oil actually fractioned out-- that is, the first fraction to be released was often the thinner, more volatile fraction. Casing-head gasoline was a tricky, highly explosive compound which was originally considered to be an unwanted occupational hazard of the drilling process. It and the natural gas were both hazards which could lead to one well or an entire field going up in roaring flames.

The Bradford Oil Basin

Early in the history of drilling, such casing-head gasoline was simply ducted off from the well and allowed to run down the local stream channels. An incredible amount of light hydrocarbon was simply allowed to go free into the environment in the early oil fields.

Nost natural gas was piped to the edges of the field and burned off, intentionally, as an avoidance of explosive combustion. Gass flares were the tricky and spectacular norm in the Bradford fields.

In bradford city itself, Natural Gas was the logical alternative to city producer gas for lighting and for heat. After about 1850, this amounted to open gas flares. These were particularly hazardous in a city of wooden buildings and large petroleum holdings, and fires were both common and catastrophic.

After 1891, most city, and many country homes were lit by natural gas mantle lamps, invented by the Austrian chemist, Carl Auer von Welsbach. The mantle gas lamps were far brighter and far safer than open jets. Bradford became known as a city to be seen by gaslight, as were London and Paris.

Eventually, natural gas would be used more and more to run pumping powers and drilling engines, a way of "sustainably" using a fuel at hand within the oil field itself.

Kerosene

In the same decades, however, the competing invention of the kerosene wick lamp, followed by the Aladdin-style kerosene mantle lamp was to revolutionize lighting, as well as the demand for petroleum products.

Kerosene was the next heavier fraction in the natural well after casing-head gasolene. It was a clearish yellow strong-smelling liquid that was dense enough

Wellsbach natural gas lamps.

Kerosene lamps.

Aladdin mantle lamps

to burn as a contact oil instead of volatilizing and exploding.

Soon kerosene was the more lucrative product, as demand matched, if not superceded the call for heaver petroleum. Thousands, and then millions of barrels of the desirable kerosene were shipped down the Allegheny river to Pittsburgh.

As natural gas was still a lucrative byproduct of the oil field, Glass factories had sprung up along the edges of the fields. Pittsburgh was recipient of numerous patents for improved kerosene lamps and burners. And glass lamps previously all the rage for whale oil were now rethought as lamp founts for petroleum instead.

Whale oil burned at the end of a small tubular wick much like a candle-- it burned with a clear, smokeless light. In contrast, kerosene required a flat wick, and even then it produced a vast amount of smoke, as well as its yellowish light. The improvement was to add a clear glass chimney to the lamp, which concentrated air flow, and forced the kerosene to burn hot, brightly, and mostly without smoke.

Initially, chimney glass was itself a danger, since the glass expanded and contracted enough to make the chimney short lived. In addition, the quality of casinghead kerosene varied from producer to producer and even well to well. Kerosene was often mixed (intentionally or unintentionally) with a certain fraction of casinghead gasoline. And a kerosene lamp filled with gasoline was potentially more of a Malotov Coctail waiting to happen rather than a safe lighting device.

The age of petroleum lighting extended the product range of the oil producer. It also made him inventive, with two particular and related developments coming out of Bradford itself.

The Bradford Oil Basin

Petroleum Refining

The first of these was the concept of specifically and difinitively separating the individual volatile fractions of the liquid petroleum.

This was first attempted successfully by Samuel Kier, who split crude oil into lamp oil in 1850. Kier had begun in the southern fields, packaging petroleum oil as patent medicine. He then marketed kerosene for lighting. The legends tell that he began his separation and purification experiments by heating crude oil on his wife's kitchen cook-stove.

He moved his experimentation to other quarters after he managed to set his wife's kitchen on fire.

Kier's observation was that when crude oil was heated, different volatiles were driven off in a particular order, the lightest fractions coming off first, and the heavier fractions remaining until higher temperatures were applied to separate them. Kier's "Carbon oil" for illumination sold for a dollar fifty in Pittsburgh and New York City.

In its simplest form, the "refinery" was a large cooking tank, heated by natural gas flame, and a set of successive pipes kept at differing temperatures to catch the various fractions.

The Bradford Oil Refinery has the distinction of being the oldest continuously operating oil refinery in the world. Its original capacity was described as being 10 barrels of oil per day. This was later expanded to an eventual capacity of 10.000 barrels per day.

It was long pointed out proudly by the older members of the Cawley family that Bradford, Pennsylvania sold oil to Saudia Arabia for some fifty years.

Kendall Refinery, Bradford, Pa.

Jon C. Cawley

Gasoline and the Carburetor

The first working carburetor was developed in europe bySamuel Morey in 1826. Later, in a direct application for automotive engine work was made by Karl Benz in Germany in 1888.

By 1897, in Bradford, George Holly and his brother Earl had already begun to design and build early motorcycles and an automobile that they called "the Runabout." Their design improvments included schemes for gasoline carburetors.

Their efforts had caught the notice of Henry Ford bythe turn of the 20th Century, and by 1903, he had asked them to design an inexpensive gasoline carburetor especially for his "Model T" Ford. The American age of Gasoline automobiles was now officially begun. The Holly carburetor was to become a powerful American classic.

Eventually, a Texas oil-man by the name of Fred Koch developed the technology to not only split petroleum into fractions, but rather to chemically "crack" it into whatever percentages of various desired fractions.

By this time, that old waste product, gasoline, with the birth of the Holly Carburetor, was now in great demand for the early automotive age.

The Zippo Lighter

New oil products spawned new uses for oil. In 1932, George G. Blaisdell attended a dance at the Bradford Country Club, and noticed the cumbersome time that patrons were having lighting their cigars and cigarettes. Blaisdell in turn used a not-so-in-demand portion of Oil "lighter fluid," a wick, a metal case, and a metal wind protection collar to invent the wind-proof Zippo lighter.

The Bradford Oil Basin

Kendall 2000 mile Motor Oil

The American love affair with the automobile was a progressive endeavor. Cars became sleeker and faster, and were built with bigger and more powerful engines. In 1928, the Kendall Refining Company in Bradford had responded with a now highly-refined version of motor crank-case oil.

The new oil was considered "pure enough to drink a spoonful" without harm. Experiments by the company showed that the new technological oil could be run in the transmission case of a standard civilian automobile for 2000 miles without harm. Kendall adopted a red, two-finger logo at this time to underscore this technological milestone.

Various "Penn-grade" oil products were developed by both technology and marketing as the 20th Century begun. One of the most successful oil marketing schemes of the era was that of the Sinclair Company, whos advertisements featured large and impressive "Brontosaurus' dinosaurs. If one looked carefully, they might notice that regular grade Sinclair motor oil was "Mellowed 100 million years", while their premium oil, from the same renineries, was advertized as being "mellowed for 300 million years!"

The pumping of oil wells, and the refining of oil into myriad inventive products increased the size and prolonged the life of the Bradford sand production--and oil demand. In 1881, Bradford was supplying 83% of the oil produced in the United States, and 77% of all the oil produced in the world.[1]

After this great peak, however, production began to slip in the Bradford region, despite shooting and pumping. Bradford wells produced over long periods of time, but at lesser and lesser rates.

From the 22,945,069 barrels produced in the "golden" year of 1881, production fell quickly to about eleven million barrels in 1885, to about six million barrels in 1890, to about three million barrels in 1895. Bradford oil field production reached a low point in 1906 at 2,022,000 barrels, and from 1906 until 1926, production remained at between two and three million barrels per year. It was painfully apparent that the Bradford field too, had its limitations.

With the decline in production, many lease-holders abandoned their wells, or sold them for a trifle of their possible value. The price of oil had dropped to match Bradford's production levels as other fields in California, Oklahoma, and Europe began to come on line. The automotive industry was just beginning, and it was not until after 1911 when the need for gasoline and motor oil overtook the market. Electric light was making its appearance with the 20th century, and the kerosene market was also on the decline.

And yet many producers realized that there was still much oil left in the sands below. The problem lay with extracting it from the rock. Bradford producers, loyal to oil business and to the region, soon began to experiment with the methods of secondary recovery.

Vacuum Recovery

One of the earliest forms of secondary recovery was to supplement well pumping with a reduced surface pressure on the well. This was done by placing vacuum pumps on the tubing at the well tops.

Within a few years it was being improved by Bradford producers. It became more popular as production rates plummeted in the 1890s. And it saw its greatest use as a secondary recovery method around Bradford between 1913 and 1921, before water flooding was legalized.

The Bradford Oil Basin

Vacuum was a logical next step to pumping, since pumping itself was based on decreasing pressure. Vacuum pumping did increase well output, and produced lighter and thinner oil than usual for any given well. This was because under lower pressures, the oil in the well tended to separate. The lighter, more volatile parts "boiled out" of the oil, and were preferentially pumped.

When the pumpings were repressurized at the surface, they contained relatively large amounts of casinghead gasoline, "refined" within the well itself. But this led to some problems with the method as well. As volatiles were preferentially pumped away, the oil at the bottom of the well became heavier and more viscous. It flowed much less easily, and the tendency for paraffins to settle out onto the sand surface and well parts was greatly increased.

The vacuum wells needed to be cleaned and coaxed often. The overall effect of vacuum pumping on Bradford was minor, and vacuum pumping was hardly a panacea for Bradford producers. Still, the individual wells would output. And even today a few vacuum systems are in use in and around Bradford.

Water Flooding and the Second Boom

Early on, some producers noticed that abandoned wells continued to show signs of oil, and that fresh water which had flooded some of the wells through leaky casing or from ground water seemed to be affecting the output of other wells.

The first recorded incident of well flooding was on Columbia Oil Company land near Oil Creek in the Southern Pennsylvania field. Abandoned wells belonging to Columbia had been allowed to rust through, and by 1876 were taking on water, flooding the Venango third sand layer.

Jon C. Cawley

Wells on the adjacent land, belonging to the partnership of Barcroft and Kirkwood, soon began to show greatly increased production. The men soon realized that it might be the water flowing into the adjoining wells that was helping. They took great precautions to keep the Columbia Company from discovering the flood, including taking away their own increased production at night. With time, Columbia found the leaky wells and reconditioned them. The Barcroft and Kirkwood wells quickly declined.

Since water was presumed to destroy the oil sands, great precautions had always been taken to keep water out of the wells. Abandoned wells were seen as a constant danger. After 1878, there were explicit state laws in Pennsylvania forbidding flooding (intentional or otherwise) and requiring that all abandoned wells be plugged. Further laws were passed in 1881, 1885, and 1891. Because of the laws and the common beliefs of the producers, people did not consider flooding as any kind of solution to the production problem.

In 1880, John F. Carll of the second Pennsylvania Geological Survey published a report calling attention to the possibility of recovery by flooding:

The flooding of an oil district is generally viewed as a great calamity, yet it may be questioned whether a large amount of oil cannot be drawn from the rocks in that way than by any other, for it is certain that all the oil cannot be drawn from the reservoir without the admission of something to take its place. If one company owned all the wells drawing upon a pool, and had accurate records of the depths and characteristics of the oil producing stratum in each well, it is quite possible that some system might be devised by which water could be let down through certain shafts, and the oil forced toward certain other shafts where the pumps were kept in motion, and the rocks be completely voided of oil and left full of water.

The Bradford Oil Basin

As it is, however, no systemized plan of action can be adopted. The careless handling of one well, by which water is let down to the oil rock, may spoil several others belonging to different parties. A clashing of interests at once arises and is likely to result in disaster to the whole district. (Carll 1880)

CarlI's report was ahead of its time, and it was little noticed or discussed by producers.

By the early 1890s, however, Bradford district producers were beginning to notice the effects of flooding on their own. The Bradford district had always been slightly lax, and some number of abandoned wells were allowed to fill with water. One of the earliest of these floods was about two miles southeast of Rew, and another about one mile north of Aiken.

As surrounding production increased, Bradford producers began to see the value of water flooding as a secondary recovery tool. Between 1900 and 1921, a small number of Bradford producers laid the roots of water flooding.

The accidental "dump" method of putting water into the wells was improved upon, and input wells were tubed to keep debris out of the water flow and from clogging the sand below. Because of the state laws on plugging and flooding, these experimental floods were done more or less secretly.

The "circle" flooding method used during this time was simple. The procedure was to take a central nonproducing well and let water into it. The water forced the oil outward in a circle from the central well toward surrounding wells. These were pumped. The procedure was hardly systematic. Very little was known about the characteristics of the oil sands, and thus surrounding wells were often spaced much too far away from the central well.

The Bradford Oil Basin

New wells were often drilled around the input as the flood reached the producing wells and they were "watered out." The circle of water was asymmetric and it was hard to predict the size and shape of the flood area as it spread. But the system seemed to work.

In 1921, in response to pressure from Bradford producers, a special measure was passed into law in Harrisburg making flooding legal in the Bradford third sand. The Act of 1921 voided many of the earlier laws.

"The flooding law recognized a practice which pioneers in secondary recovery had followed for some years despite the plugging laws in effect." (Murphy 1948)

It was supplemented in 1923 by an act allowing flooding in the Bradford, Kane, and Haskill sands. Another act in 1929 extended this to most of the other oil sands.

The Forest Oil Company has been identified with water flooding almost from its inception, and helped win for Bradford the distinction as the inventors of the flooding process. In about 1916, Forest Dale Dorn and his father Clayton Glenville Dorn incorporated Forest Oil and began to study and experiment with water flooding.

At a time when the Appalachian oil fields were being cited as a prime example of resource squandering and exhaustion, Forest Dorn was becoming increasingly confident of the possibilities for secondary recovery. The company bought land and holdings in the Bradford fields.

When flooding became legal in 1921, Dorn's ideas were quickly adopted throughout the industry, and the company took a leading position in the pioneer field.

Forest Oil's waterflood production grew from its 1916 rate of thirty-eight barrels a day, to over 9300 barrels of oil a day in 1939. This amounted to more than 20% of the entire Bradford field production, and made Forest Oil the overall largest single producer of Pennsylvania crude oil.

In 1922, Forest Dorn developed a better alternative to the circle flood. This new "line flood" consisted of a row of water intake wells offset on either side by rows of pumping wells. Water was let into the center row of wells, and the outer wells were pumped. The oil was pushed outward to the pumping wells from the entire center area.

When the outside row of wells had reached its economic limits, another row of pumping wells was drilled further outside, and the old row was converted to water intake wells. This was more efficient than the circle flood, and its boundaries could more easily be calculated and controlled.

With water flooding, Bradford production began to increase, but slowly. In 1923, State Geologist George Ashley stated:

"Little news has come out of the Bradford-Allegheny [sic] field during the year. Apparently work in those fields has just about kept pace with that of earlier years. Taking September as an average month, Allegheny [sic] County, New York, shows twenty-five completed wells with a production of seventy-five barrels, or three barrels per well, one dry hole and one gas well. Bradford County [sic], Pennsylvania, shows sixty-four wells completed, all oil producing, average yield about three barrels. This field still maintains its remarkable record for percentage of producing wells." (Ashley, 1923)

Bradford was hardly a newsworthy item in 1923. Its annual output stood at a little less than three million barrels for the year. By 1926, however, the same report read:

The Bradford Oil Basin

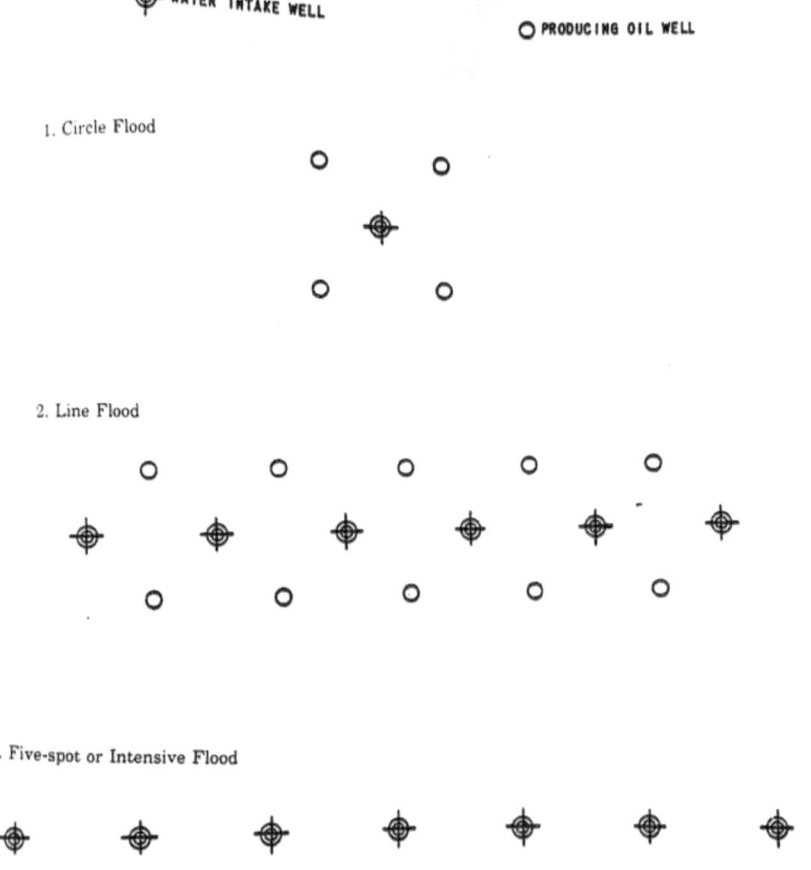

⊕ WATER INTAKE WELL   ○ PRODUCING OIL WELL

1. Circle Flood

2. Line Flood

3. Five-spot or Intensive Flood

Jon C. Cawley

"In the Bradford district of Pennsylvania, and the Allegany field, development work has been pushed to an extent that has not been known for many years and production kept constantly increasing. Although the demand for Pennsylvania crude has been increasing in recent years and although most of the Western Pennsylvania refiners have enlarged their capacity during the past year or are now doing so, the situation in this part of the field is liable to become acute any time. Production is increasing so swiftly that the time may come when the price differential between the Bradford grade and the other Pennsylvania grades will be wiped out after production exceeds consumption. Bradford oil at present brings ten cents more than that produced in other sections." (Turnbull 1926)

Bradford production in 1926 stood at a little more than 4,700,000 barrels.25 In 1928, water flooding was again improved. In the new "five-spot" or intensive flooding system, the rows of wells were simply reversed. The input wells were on the outside rows, and the pumping wells were in the middle. This way the oil moved toward the center well from all four sides. This resulted in higher and faster recoveries.

Frank Haskell of the Associated Producers Company is credited with the theory of five-spot recovery. Associated Producers had been experimenting with the system, but with few results because of the spacings they were using and the low permeability of the sand they were trying to flood.

The first really successful five-spot flood was by Arthur E. Yahn of Olean on the Kuno-Kuhn lease in the northern part of the Bradford field. His results were good, and the five-spot method soon caught on.

At about the same time pressure flooding was introduced. Pressure flooding supplemented the capillary action of flooding by putting water into the wells under

The Bradford Oil Basin

high pressure. The oil was then literally pushed through the sand toward the pumping well. Pressuring was begun in the Bradford fields in 1925 and 1926, but the first significant output was in 1928 by John Messer of Bolivar, New York. Messer's production was up, and the pressuring method was quickly applied throughout the region.

The new technologies and increased output of the late 1920s began a new approach to oil production. Emphasis was placed on scientific analysis of the oil fields. Newer and stronger pumps had to be developed to handle the excess water that came with flooding. The porosity and permeability of the oil sands needed to be carefully measured and considered. The various sand layers needed to be correlated.

Soon the Pennsylvania State College was allied with the producers of Bradford to work on the problems of the Pennsylvania oil fields. Initially the major producer groups - Bradford District of the Pennsylvania Oil Producers Association, Pennsylvania Grade Crude Oil Association, Pennsylvania Natural Gas Men's Association, and the South Penn Oil Company - sponsored several annual research conferences at the College. The first of these was held in October, 1930. It was hosted by the Penn State Dean of Mineral Industries, Edward Steidle, and was the first major conference held in the college's new Mineral Industries Building (now Steidle Building).

Additional support from the Bradford District of the Pennsylvania Oil Producers Association and from the state of Pennsylvania enabled Penn State to start a major secondary recovery laboratory and a program of petroleum and industrial research. This Petroleum Experiment Station was important to Pennsylvania production technology for many years until it finally gave birth to Penn State's present Petroleum and Natural Gas Department in the late 1950s.

The Bradford Oil Basin

Jon C. Cawley

Meanwhile, the Bradford basin was diamond drill cored. Permeability and porosity profiles were developed and the region continued to be drilled. Production rose steadily. In 1930, Bradford produced about nine million barrels of oil. In 1934, the total was just under twelve million barrels. And in 1937, a new peak of production was reached at 16,738,688 barrels of oil.

In that year, 4111 wells were drilled in the Bradford region, more than were drilled in the year 1880 at the peak of the first great boom. 1937 was a prodigious year for a field that had been declared dead two decades before. It is interesting to note that the new frugal Bradford producers put much of their new money directly back into oil research that might hold their industry stable in the future.

After 1937, however, another long decline did soon begin. This time the production drop was not nearly so much a factor of supply but more so of demand. The discovery of other great oil fields throughout the world had made gasoline available for the new automobile age, and much of the Pennsylvania grade crude was being refined as lubricating oil.

By the late 1930s, the market was feeling the effects of the Great Depression. The high Bradford production came at a national low in production and commerce. Bradford, it had to be realized, was a second generation field. And although Bradford was of good size for its time, it was no longer the archetype of the oil field giant. It had already been surpassed by the fields of the Gulf Coast and the Middle East. It no longer controlled the market. Instead, it was finally beginning to be controlled. As C.L. Suhr of Pennzoil stated in a 1940 morale address to local producers:

"Many changes have been witnessed since the early years, outstanding among which probably are the changed position of our Pennsylvania industry as regards its percentage of total output and its share of total markets. So much additional Pennsylvania oil has through secondary recovery methods been made available [that] when the very survival of our small wells under the weight of ever increasing competition and the tremendous good that has come from the ability to recover that difficult portion of our reservoir crude considered, there is indeed much to be thankful for." (Suhr 1940)

But the producers were losing money. By September of 1940, the price of oil was lower than the overall cost of production for many Bradfordians. Many producers during this time actually lost money with each barrel of crude that they pumped. The Bradford region dug in and waited for better times.

Jon C. Cawley

The Bradford Oil Basin

Jon C. Cawley

Chapter 9
Bradford after 1941

World War II

A better market came with the second world war, but again, this change did not quite come as expected. For the first few months of the war, Bradford producers held their breath, waiting for the much needed upswing in oil prices. At first it did not come:

"This was a funny kind of war. For a long time there was more war on the radio than there was in the trenches. They simply weren't using as much oil as the buyers had expected." (C.L. Suhr)

And then when the upswing did come, it came quickly, and the wounded production of Bradford could hardly keep pace. From 1943 to 1948, the new wells drilled in the Bradford district averaged over 2000 per year-and Bradford production only remained fairly constant.

During the war years, the Pennsylvania Grade Crude Oil Association joined the Penn State Research Station. In 1944 they, with several other industries, raised a fund of $300,000 to support three years of research for better oil recovery. In 1945, Penn Grade established a comparable recovery laboratory at Bradford itself.

The Bradford Oil Basin

The lab was equipped to study pressuring, chemical plugging, and brine recovery. Group research over the next three year period was split between the Penn Grade Laboratory, the Penn State Research Station, Battelle Memorial Institute, St. Bonaventure College (at Allegany, New York), and the U.S. Bureau of Mines at Franklin, Pennsylvarna. The Penn Grade Laboratory was eventually discontinued in 1955.

One of the major contributions to come out of the war years research was introduction around Bradford of produced water for flooding. In the early years of flooding, Bradford always relied upon "raw" surface or shallow well water for flooding. There were three problems with this. First, raw water tended to be hard, containing iron and manganese as well as calcium carbonate. These tended to settle out with pH changes in the well, and would seal off the surface of the sand. Sometimes the impurities were even more obvious and several early floods were unsuccessful because of silt and clay carried into the intake wells with the water. Such impurities needed to be removed.

A second problem came from algae growth. Raw water put into input wells quickly produced great masses of algal slime which clogged pipes and in some cases decomposed the pipes or oil. When the water was warmer than usual, and light was available (as in holding tanks), blue-green algae grew abundantly. In anoxic, dark conditions in the presence of oil and iron pipes, a corrosive "black algae" flourished. And when the water was cold and oil was present, a white algae growth was formed. Produced water needed to be treated with a bacteriacide/algaecide.

And the third problem came from oxygen dissolved within raw water. Because the oil wells were naturally in a reduced condition, the presence of oxygen greatly increased corrosion problems in the system. Produced water had to be deoxygenated.

BR-51—Hotel Emery and Music Stand, Public Square, Bradford, Pa.

The Bradford Oil Basin

BRADFORD, PA.   Bradford City Park and Post Office

CARNEGIE LIBRARY, BRADFORD, PA.

Jon C. Cawley

One of the first plants in the Bradford region to use produced water as its main supply belonged to the Quaker State Refining Company. It was built about 1950 at Bell's Camp near Derrick City, and served as a prototype for many other plants in the region. At the plant, water was chlorinated (to kill algae), softened (to precipitate iron and manganese), and limed (to neutralize corrosives). Coagulents were added and the water went to large settling tanks to remove debris. The water was then filtered, and went to the wells.

If water put into the floods was carefully controlled, the water coming back out of the floods was less so. By the 1950s there were tens of thousands of holes drilled in McKean County alone, and soon much of this area was placed under flood. The Bradford sands were no longer a closed system. As areas were watered out, the flood often found its way to the surface through old, unplugged, and abandoned wells.

In the most obvious cases, these forced-artesians carried oil-tainted water into the surface environment. This was messy and unsightly, but did little actual harm. The Bradford oil was relatively inert, and soon removed itself from the active environment as it had after the boom year spills of the 1800s.

More often, however, the algae-coated casings carried a thin oil-sand brine containing sulfides, oxides, and various mineral salts from below. This used water ran into many of the local streams, and caused more repressive environmental problems. The mineral rich waste water effectively removed much of the aquatic insect population within the streams. With the base of the food chain disrupted, trout and other stream life began to disappear as well. This sort of water quality problem within the region was little studied.

In 1955, only three hundred seventeen wells were drilled in the Bradford region. Production dropped that year, and again markedly in 1957. The field was again at

The Bradford Oil Basin

a low. Still, oil was being produced, and the Bradford district was supplying over one-third of the oil produced in Pennsylvania. This oil was primarily refined locally by Quaker State, Kendall, and Pennzoil.

As the 1950s and 60s progressed, oil production around Bradford continued to decline, but slowly. Mechanical logging was introduced, and well fracturing by hydrostatics finally replaced well shooting. The local oil industry was stable, and producers accepted the long decline as part of the business. The period between 1950 and 1970 was quiet around Bradford. The petroleum laboratories were disbanded in 1955, to be replaced later by a petroleum program at the new Bradford branch of the University of Pittsburgh.

In 1973, the Arab oil embargo raised oil prices, and sparked some new interest within the Bradford field. New drilling increased, and many older wells were reconditioned. Output was still relatively small, but was again stable as compared to the market. Portions of the Bradford third reserves were tapped, and some wildcatting was begun in the marginal areas around the main Bradford field.

Some producers of the 1970s turned attention toward natural gas, and toward oil and gas potential within the deeper strata under Bradford. An optimistic state report in 1970 read:

"A vast, deep petroleum potential, representing 83% of the total prospective stratigraphic section, remains untested in Pennsylvania. Beneath the heavily drilled Devonian oil and gas belts of the Plateaus Province (sic) lies from 5000 to more than 20000 feet of Cambrian, Ordovician, and Silurian strata which contain less than two wells per twenty-five square miles The present and future need for hydrocarbons combined with proximity to expanding industrial areas should provide the incentive for rebirth of the industry in Pennsylvania. The deep challenge awaits and will be met if mod-

LADIES CLUB, BRADFORD, PA.

The Bradford Oil Basin

HOTEL HOLLEY, BRADFORD, PA.
BRADFORD'S ONLY FIREPROOF HOTEL

Second Ward Grammar School, Bradford, Pa.

ern technology is utilized and a cooperative environment is established which encourages joint ventures to divide the financial burden of deep drilling. (Kelley et al. 1970)

In 1986, it is still too early to tell the full extent of this deep drilling potential in the field.

Tertiary Recovery

Also in the 1970s came a new interest in tertiary recovery. By 1974, threefold and fourfold increases in the price of oil once again made it economic to produce oil around Bradford. Producers soon began new programs in recovery methods. As major portions of the Bradford pool had been watered out in the 1950s and 1960s, they were effectively removed from production. But still the sands contained more oil than water.14 The oil that was left surrounded sand grains, and adhered to clays and feldspars. The adhesion of this oil was more than water flooding, even pressure flooding, could break.

Many ideas have been advanced as to how to produce this oil, including well injection of detergents, brines, and polymers. The major concern has been with cost, but there is also still the problem of the injected fluids reaching the surface, or polluting groundwater supplies. In 1975, Pennzoil, Witco Chemical, Quaker State, and the U.S. Department of Energy began a joint project at Derrick City. The Penn Grade Micellar Displacement Project consisted of twenty-four acres of polymer injected wells. The polymers used were intended to displace water and help force out oil remaining in the sands. By 1980, the project was well established. The land was showing slight increases in oil production, and a lessening of water from the wells.

Other projects of this sort have included a similar "Maraflood" project at Cyclone, and a microemulsion-flood project by Pennzoil (the Emerson Lease Project). A

The Bradford Oil Basin

second sort of enhanced recovery has been the use of brines (actually mineral-rich oilfield waters) to stimulate oil and natural gas production. By the 1980s, there was some interest and debate on the practicality of brine fracking methods in gas and oil recovery.

In 1984, the Benson and Reynolds Gas Company of Bolivar, New York, began a brine recovery project for natural gas in Potter County, Pennsylvania. The project used excess well brines produced in adjacent fields. The program caused much public concern over local water quality and eventual responsibility over the injected test fields. By 1985, the company had applied to the Environmental Protection Agency for permits to carry on with portions of the project, suggesting that the amount of gas produced would be significant and valuable to produce. The EPA granted final permission by the end of the year, and the project is ongoing.

With high oil prices continuing into the 1980s, local production remained stable. McKean County production stood at 887,904 barrels in 1980, and Bradford oil averaged $38 per barrel. Oil research and development continued to spread in the Bradford region, and although the Bradford sands are overdrilled, tired, and watered out, there still has been some local optimism over the future of Bradford production. Small producers have continued to drill new wells in the outlying corners of the field. Old wells have been reconditioned, and production has continued, although at nearly subsistence levels. The local refining companies - Quaker State, Pennzoil, and Kendall (now Witco Amalie) have continued to produce Penn Grade oil products from Bradford crude.

By the 1970s and 1980s, the EPA was playing a major role in the Bradford region. After a hundred year history of laissez-faire policy on oil and drilling in the oil-soaked Bradford district, there was now an increased concern over oil leaking into the environment. A well-meaning attempt by the federal agency to save the century-old oil field from polluting itself ("If I

The Bradford Oil Basin

Above: Bradford Regional Airport
Below: Corning Glass Works
    (Bradford Electronics Division)

have a sheen on the surface of the waters of the United States, I have an emergency." -Roger Meyer, EPA on-site coordinator), has added yet another burden to the small producers of the Bradford area.

In a major action in June of 1985, EPA officials, headed by Roger Meyer, legally declared most of the northwestern Pennsylvania oil fields as an inland oil spill. A major cleanup effort was begun in the summer-combined effort of the EPA and the Coast Guard Atlantic Strike Force. The project centered initially in the Allegheny National Forest lands, and in the waters of the Kinzua Dam. By November of 1985, the EPA had cleaned up one hundred fifty-three spills in the Warren area forest region.

In one instance, oil from a leaking well had streamed down thirty yards of hillside, according to records. Tons of contaminated soil and an estimated 78,000 gallons of oil were removed from the [various] sites and boulders coated with the yellow, paraffin-based oil of the region were steam cleaned.

"Although we got a lot of flak for steaming the rocks, it was cheaper than carting them away," said Vincent Zenone, an EPA technical consultant.

The EPA has begun negotiating with the (oil) company to recover the $524,000 it spent on cleanup costs, agency officials said. (Philadelphia Inquirer Nov.1985)

The work should spread to other areas with time.

Great public concern about the new program has come from both the local oil companies and private citizens. First, the cleanup effort deals only with oil, and overlooks brine contamination, which is likely to be the more serious environmental problem. Second, many see the "emergency" cleanup of a problem that has developed over a hundred years as "tantamount to overkill."

The Bradford Oil Basin

And finally, the EPA policy (to date) is to charge the owners of the spilled areas mineral rights or land. Because EPA standard cleanups are expensive, the large bills may well be fatal to many of the marginally profitable small companies or single producers. Many of the spill areas are likely to be old unplugged wells from the early years, on land now used for other purposes.

Billing is likely to affect not only oil producers, but also farmers, lumber companies, or other landowners who have never seen even marginal profits from their casings. It is a serious problem that will surely have an effect on the future of the oil regions.

In the early months of 1986, the Middle East OPEC began to fail apart. With import trends broken, the price of oil quickly began to slip. From a January rate of $38 per barrel, the market rate was down to $14 per barrel in April.

Although American motorists are celebrating a radical fall in gasoline prices, the American oil industry is once again on unstable ground. And presumably the Bradford fields are again less than economic. Once again Bradford producers are digging in for hard times. As time goes on, tertiary recovery and deep drilling projects will be put on hold. Small producers will leave their wells to sit, and many leases will be lost to disuse and paraffin. Some producers are optimistic that the prices will rebound. Others feel that the Age of Bradford is indeed finally over.

Above: Kinzua Seneca Treaty town along the Allegheny River.
Below: Kinzua Dam

The Bradford Oil Basin

Jon Clayton Cawley

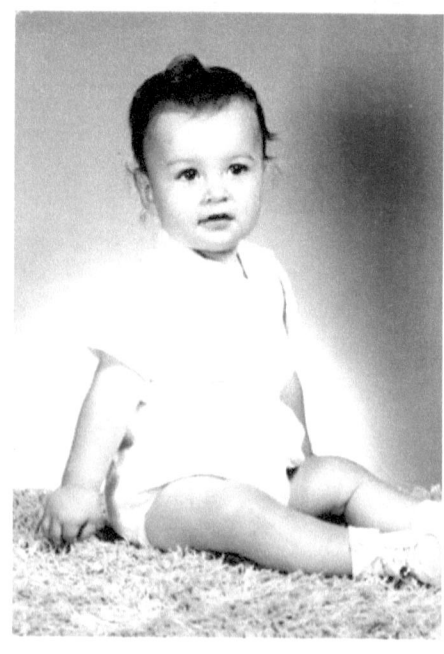

Jon C. Cawley

Chapter 10
Geology of the Bradford Field

Beneath the forests and oil pumps of the Bradford district lies the geology of the Allegheny High Plateau. This Plateau Province is a region of relatively flat hills cut by deep valleys. The plateau extends from the Great Lakes in the Northwest, to the Ridge and Valley Province to the Southeast. Bradford is situated near the middle of the region, and just beyond the southmost extent of the most recent (Wisconsin) glacial advance. The stream channels of the northern Bradford fields are filled with glacial gravel. The stream courses cut older rocks of Devonian to Pennsylvanian age, some two hundred and fifty million years old.

The rocks are mostly various sandstones and shales. Small coal beds exist, and there are minor limestones and clay beds as well. The hardest, most resistant shales and sandstones form the high plateau ridges of the region, and softer shales and sandstones are exposed on the slopes and in the valleys. These flat-lying rocks are the sediments of an inland (geosynclinal) sea. The rock outcrops of the eroded hills are ancient sea floor sediments and delta deposits. The oil sands far below are beach and bar deposits of an ancient ocean bay.

Streams of the Allegheny River Basin cut slowly into the rocks of the Bradford district; only the youngest rock layers of the region are exposed. The steep hillsides are covered with thick soils and forest land. Rock outcrops and geological contacts are not always easy to find; much information comes from soils and roadcuts, from stream gravels, from outcrops and small local quarries. Information on deeper geology comes from well cores, and from the drillers' log books of a hundred years of oil drilling.

Deep wells, sunk in search of natural gas, extend this record even further, down to the Queenston Redbeds of the upper Ordovician time. Using this knowledge, it is possible to describe and interpret much of Bradford's geological history.

Soils of the Bradford Region

Overlying much of the Bradford region are the local soils formed by the chemical and physical weathering of parent rock materials. The amount of weathering and the characteristics of any soil depend on the nature of the parent rock, the type of climate, the relief, the plant and animal life, and the length of time these factors have influenced development. These soils are the youngest geological feature of the basin, and so will be considered first.

There are five major soil associations in the Bradford region based on topography, the parent rock material, and soil texture and composition. These group divisions around Bradford are tied in closely with land use, and represent patterns of soils rather than specific soil types.[3]

The Cavode-Gilpin-Wharton association includes the soils of the high ridgetops. These are sandy, and somewhat infertile. They cover land used for limited farming, and land left as forest. The Cooksport-Hazleton-Gilpin association is stony valley-side soils, and covers much of the Bradford region. These are acid soils, stony, and of questionable fertility---they are mostly devoted to forestland.

Jon C. Cawley

The Bradford Oil Basin

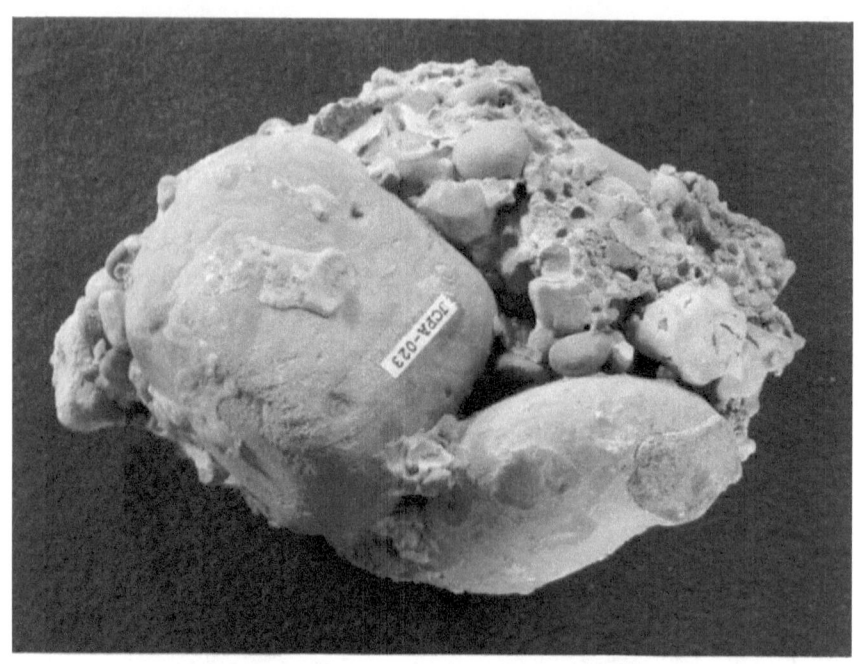

The Gilpin-Hazleton-Ernest association is also stony valley and slope soils. These are more sticky, clay soils which tend to stay wet late into the spring. They are sometimes farmed. The Leck Kill-Albright-Oquaga association consists of soils in the northern section of the basin derived mostly from the exposed slope and valley shales. These soils are somewhat acid, and are used primarily for pastureland.

Finally, the Pope-Philo-Albright association includes the floodplain soils of the stream and river valleys. The level flatlands provide much of what agricultural land is in cultivation in McKean County.

Glacial Deposits

The next oldest geologic deposits of the Bradford region are the glacial outwash gravels from the ice sheets of the Pleistocene Age. The glaciers of the last (Wisconsin) ice age, some ten thousand years ago extended only to the northwestern corner of the Bradford basin ---- as far as Salamanca, New York. Ice surrounded the Bradford region on three sides, but never actually overrode it.

As ice built up to the north, the north-flowing streams of the basin were dammed into lakes. During this time, glacial ice obstructed the valleys of streams, including the Allegheny River, Oswayo Creek, Potato Creek, Marvin Creek, and Tunungwant Creek. The course of the Allegheny River was changed entirely by the damming--- from its original path northward into the St. Lawrence drainage system to its present course to the south. The river's great sweeping bend at Genessee, New York points it southward over the old divide, and into the Ohio River drainage system.

The gravels left behind fill the stream valleys and form terrace systems. The gravel deposits are deep - two hundred fifty feet in places - with the streams

The Bradford Oil Basin

only beginning to cut down through them from the top. The gravels are an important groundwater source. The gravel clasts are composed almost entirely of sedimentary rock types. In some places the clasts are held together by later calcite deposits, forming a fragile conglomerate.

The Pennsylvanian System   (320 to 286 m.y.)

The youngest rocks outcropping in the Bradford district belong to the Pottsville group of the Pennsylvanian Age. These consist of the Kinzua Creek formation, the Sharon formation, and the Olean Conglomerate. These rocks are relatively resistant, and are found atop the high plateau ridges of the region.

The highest formation of the Pennsylvanian rocks in the region is the Mercer Shale formation. This is a mixed gray shale containing clay beds, concretions, and some layers of coal. The Mercer formation outcrops south of Bradford in the area around Mount Alton----its coals have been named the Alton coal beds. These beds have been mined on small scale at various times in the past. Only about fifty feet of the Mercer remains as part of the Bradford stratigraphy.

The Mercer formation represents a marginally successful coal swamp of the Pennsylvanian Age. This swamp would have been the last access Bradford had to the Paleozoic inland sea system before it was lifted into a high plateau. The environment was primarily sediment controlled, with incoming silts and clays from sluggish streams overwhelming the coal plants periodically.

Below the Mercer is the Kinzua Creek formation, or the Connoquenessing Sandstone. This rock unit is a coarse-grained, nearly pure quartz sandstone which outcrops as the rimrock of Kinzua Creek in Warren County (hence its name by C.A. Ashburner in 1880).It is also plentiful to the south of Bradford on the high plateau

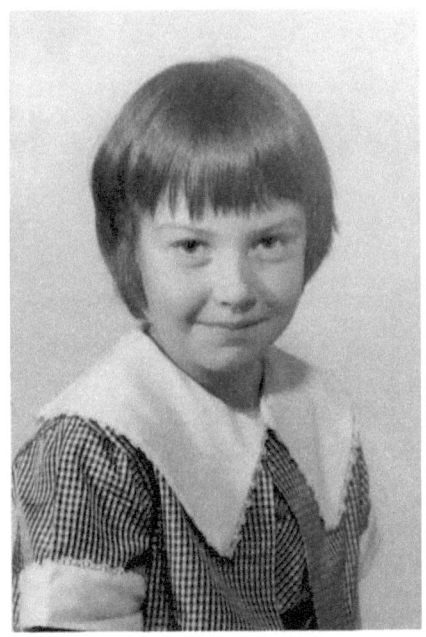

Lois Catherine Harrison

The Bradford Oil Basin

elevations. In some portions the rock is conglomerate, containing small flattened quartz pebbles. These pebbles are smaller and less abundant than those of the Olean Conglomerate below.

The Kinzua Creek formation is interpreted as fresh water sands brought into the basin by downcutting streams on uplifted highlands to the northeast. Energetic streams transported the cobbles of the Olean Conglomerate. These same streams were less violent as they deposited the sands of the Kinzua Creek, and were least active during the time the silts and coal swamps of the Mercer were accumulating.

Underneath the Kinzua Creek sandstones is the second shale unit, the Sharon formation. These shales are similar to the Mercer shales and contain one or two beds of very low grade coal. The coal is named the Upper Marshburg; it is a shaly cannel coal that is not economic to mine. The Sharon beds contain thin layers of sand as well and these obscure contacts between the Sharon and the Kinzua Creek formations. The Sharon is perhaps fifty feet thick, and represents an earlier calm period in sediment conditions during the Pennsylvanian deposition.

Perhaps the most recognizable rock unit exposed around Bradford is the Olean Conglomerate. The Olean is a coarse "pudding stone" conglomerate of white, rounded quartz pebbles in quartz sandstone. The white pebbles range to three inches in diameter, and often have a light yellowish-brown staining of iron. Many of the clasts show minute surface growths of secondary silica giving the pebbles an etched appearance. The conglomerate is resistant, and forms a caprock on many of the ridges around Bradford.

Where it is exposed, it weathers along joint lines, and forms great boulders and crevasses. The best exposure is at its type-locality at Rock City, south of Olean, New York. The outcrop at Rock City represents the highest member of the New York Paleozoic Rock Series. The conglomerate layer is from sixty-four to seventy feet thick in the areas where it is not eroded. Toward the south of the basin it becomes finer and grades into a coarse, cross-bedded sandstone similar to the Kinzua Creek.

The Olean consists of alluvial fan deposits laid down on an eroded surface of older Mississippian-aged rocks. The Olean sediments were deposited by quickly moving streams from young highlands to the north; the pebbles are well-worn and distinctly rounded rather than flat. Cross-bedding in the rock suggests an especially active deposition, and in some places a lack of imbrication, or layering, indicates almost violent emplacement.

The Mississippian System (360 to 325 m.y.)

The Olean Conglomerate rests on an unconformity, an ancient eroded surface, that marks the break between Pennsylvanian and underlying rocks of the Pocono Group in the Bradford field. The older rocks below the break were tilted somewhat to the south and then were eroded smooth (rocks layers lying next to the Olean in the north of the basin are somewhat older than the rocks in the same position to the south). The break represents a time when the region was above sea level, and was actively being downcut. In all, hundreds of feet of rock have been worn away from this surface before the Olean was laid down.

There has traditionally been some dispute as to the age of the underlying Knapp formation. This resistant group consists of two layers of gritty conglomerate sandstone divided by a fossiliferous shale layer. The entire formation is about fifty feet thick, becoming much thinner to the north. The conglomerate sandstones are iron-rich, and contain small disc-shaped jasper and quartz pebbles quite different than the smooth stones

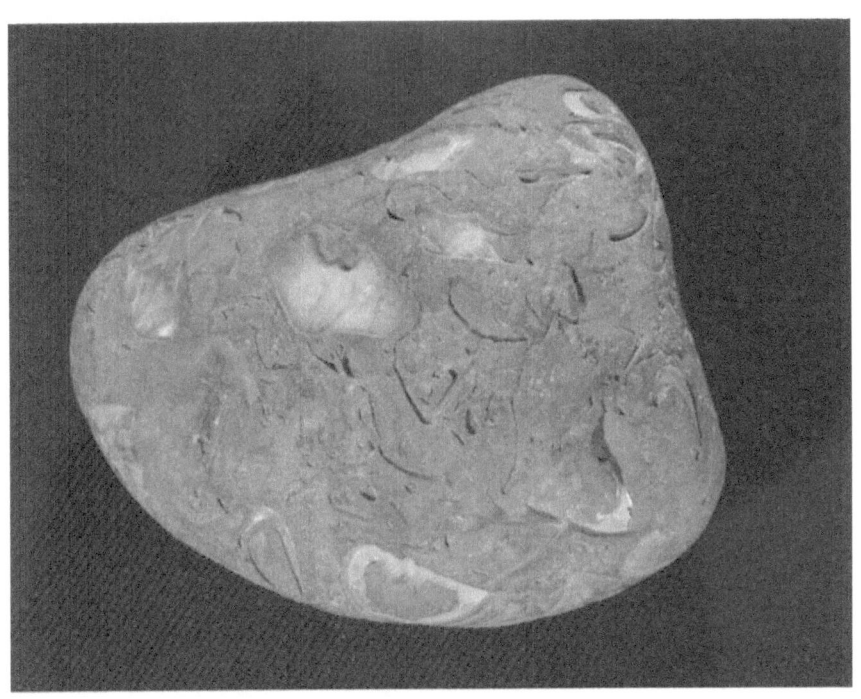

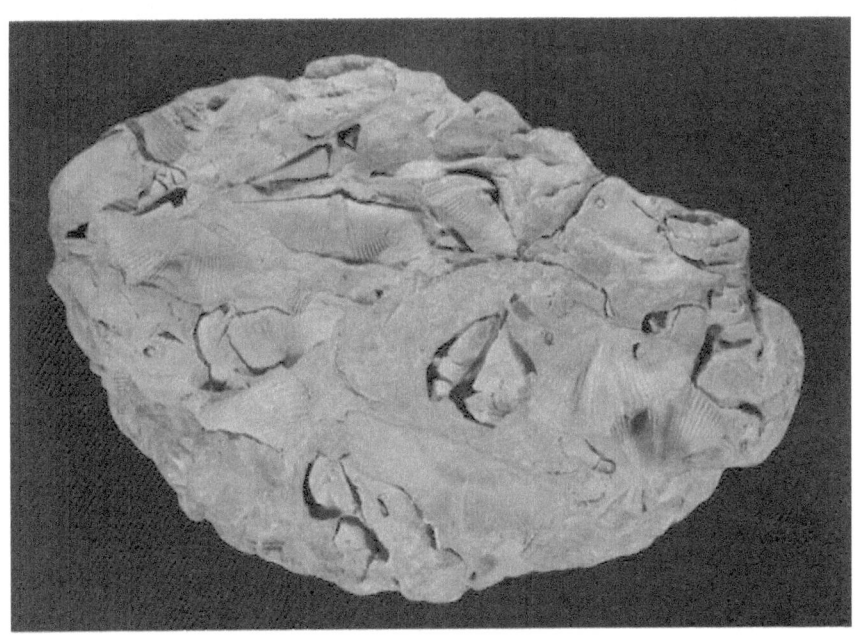

of the Olean or Kinzua Creek. They also contain marine shell fossils. The shale between the conglomerate is dark, irregular, and fossiliferous. The fossils of the Knapp include various brachiopods, pelecypods, and corals (?). The faunas are primarily Mississippian, and the age of the formation has been set on this basis. To the east of Tunungwant Creek, the Knapp becomes more shaly and grades into the Oswayo below. The true base of the Knapp formation would have to be defined by a change in fossil fauna between Carbonic and Devonic forms.

The Knapp sandstones represent the seabottom of the narrow geosynclinal sea covering Pennsylvania during the Mississippian time. The fining of the sediments to the east suggests that the Bradford region was near the northwest shore of a low landmass located in the Great Lakes area. The brachiopods and pelecypods suggest a muddy, normal marine environment.

As the area was uplifted slowly above sea level, the layers were tilted to the south and were folded slightly. They were then eroded, and much of the Mississippian deposits were lost.

The Devonian System   (417 to 354 m.y.)

Below the Knapp formation, covering nearly all of the Bradford basin, is the Oswayo formation and the rocks of the Conewango group. The Oswayo formation is composed of one hundred eight to two hundred sixty feet of greenish-gray marine shales and siltstones. The rock contains frequent sandy layers and fossiliferous beds (primarily of the Devonian brachiopod Cyrtospirifer or Spirifer disjunctus).

These nearly uniform olive shales are named after the Oswayo Creek to the north of the Bradford basin in Allegany County, New York. Toward the bottom of the Oswayo formation there is a notable fossil limestone layer from several inches to two feet thick, composed almost entirely of brachiopods (including Spirifer disjunctus and Camaretoechia).

The Bradford Oil Basin

Various stratigraphic listings note a single thick bed, or more than one thinner bed. The shell layer is probably at least partially lensatic in form. This fossil layer has been named the Marvin Creek Limestone after one region where it outcrops well. The layer is surfaced throughout the Bradford region, including particularly good exposures at the head of Indian Creek, near Eldred. The Oswayo beds are marine siltstones, shell layers, and shales of the Devonian sea. The greenish-gray shales represent an adjustment period after the Catskill Deltas from the southwest had reached their maximum extent (represented in the Bradford area by the Cattaraugus beds). The environment would have been murky and normal marine. High sediment and low lighting would have presumably limited fauna to the hardy brachiopods, pelecypods, and a few horn corals.

Before the Oswayo, came the Cattaraugus formation----some three hundred forty feet of micaceous shale and shaly sandstones. These correspond with the great Catskill redbeds and delta deposits of the south and east. The Cattaraugus shales are marine, but show much of the red coloration associated with the Catskill; the contact between the Oswayo and the Cattaraugus is defined around Bradford as being at the top of the highest bed of red shale. The upper part of the Cattaraugus consists of brick-red and grayish-green shales, with very fine-grained sandstones and some greenish-gray micaceous siltstone. The lower part is composed of dark gray to dark purplish-red shale with greenish-gray sandstones and siltstones. The rock contains abundant marine fossils, which are "typical' Catskill fauna. The upper portion of the Cattaraugus is well exposed to the west as the rocks forming the valley walls at the Kinzua Dam in Warren County. The units become coarser and redder to the south and east across the Bradford basin.

The beds of the Cattaraugus are marine pro-delta and delta face deposits. During the Devonian, the Acadian uplift to the east sent the great Catskill deltas westward into the shallow Pennsylvania sea.

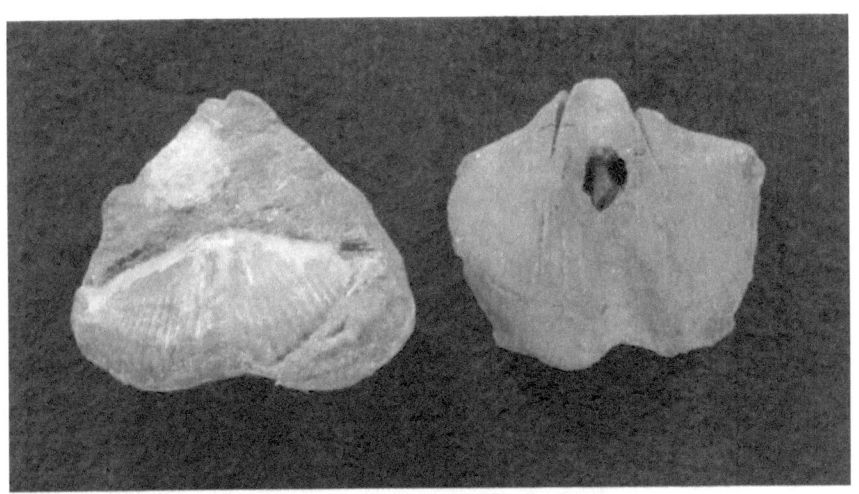

The Bradford Oil Basin

Lois C. Graham
Mary M. (Cawley) Todd
Caroline J. Crain
Jon C. Cawley

Caroline J. Crain
Mary M. (Cawley) Todd
Derian C. Ng

Jon C. Cawley

The delta sediments, retaining some amount of their red oxidized coloration, fanned out across the sea bottom in advance of the delta slopes and faces. Coarser sediments settled out close to the delta faces, and finer, more marine material extended further outward toward the west and north. The coarser sand layers probably represent short periods of greater sedimentation and turbulence.

The sandstones within the Cattaraugus are the first which are important as individual sand layers. The original Venango oil sands of the Southern Pennsylvania oil fields were sand beds of the Cattaraugus. In Bradford, the Venango sands are considered as separate conglomerate layers. The Tuna (or Killbuck) Conglomerate of the upper Cattaraugus corresponds to the Venango first sand. The Salamanca Conglomerate of the middle beds corresponds to the Venango second sand. And the Wolf Creek Conglomerate is equivalent to the Venango third sand.

The Wolf Creek forms the basal member of the Cattaraugus at its type-locality near Olean, New York. Between the Salamanca and Wolf Creek members at Bradford there is a fourth sand layer. This has been named the Hanley Sandstone because it outcrops in the old Foster Quarry of Bradford's Hanley Brick Company. The sandstone is dark red, and is of medium texture. It contains marine fossils. The Hanley Sandstone is considered to be a local development within the brick shale beds.

At the bottom of the Cattaraugus, the lowest of the red-colored layer makes a fairly continuous marker bed across the Bradford basin. The bottom of this bed has been designated as the division between the Conewango rock group above and the Chemung (or Chautauquan) group below.

The rocks of the Chemung group are primarily earlier, marine sediment versions of the Cattaraugus rocks. The unit has the positive distinction of carrying

The Bradford Oil Basin

the major oil sands of the Bradford region. Most of the Chemung rocks lie below the surface in Bradford. Only a small, uppermost portion is exposed along the banks of the Tunungwant Creek.

The Chemung rocks are soft greenish-brown micaceous and sandy shales. These shale layers alternate with many thin beds of chocolate brown compact sandstones. These sands represent beach and bar deposits, sorted sediment areas of the shallow sea. The sandstone layers become thicker and more persistent with depth to about 1500 feet.

The beds contain brachiopod fossils, many of which still include original shell material. The darker sands are distinctive because of the contrast between dark brown rock and pearly white shell. These darkest and thickest beds are the reservoir sands of the Bradford oil field.

The Bradford Oil Sands

The major oil and gas bearing sands lie within the middle region of the Chemung below a "pink rock" marker region recognized by some local drillers. There are six major sands in the Bradford basin, including the prodigious Bradford third. The sand units are usually between twenty and sixty feet thick, and vary over the region.

The first sand of note is the Bradford first, or the Glade sand. The rock is fine-grained, greenish gray, and lies about 650 feet below the top of the Chemung. To the north, the Bradford first is known as the Cuba Sandstone. It is this layer which outcrops as the old Seneca oil spring at Cuba. Production from the Bradford first is limited around Bradford, but the layer has been more productive further to the west at Warren and Tidiote, Pennsylvania.

The Bradford Oil Basin

London, 1988

Jon C. Cawley

About 350 feet below the Bradford first is the Bradford second sand, also called the Cooper or the Sheffield. The rock texture here is similar to that of the Bradford first, but contains some interbedded dark shales. The layer is about fifty feet thick where productive, and consists of two sandstone layers divided by ten feet of the shale. The Bradford second sand is very well cemented and very hard; many drilling problems and fishing jobs around Bradford were the result of a tight place in the hole at this horizon.

At a depth of about one thousand feet from the surface of the Chemung lies the most productive oil sand of the region, the Bradford third. The Bradford third sand is a very dark oil sand with brachiopod shell distributed throughout. Some fragments of carbonized plant remains having also been found, presumably transported to the sand environment from land.

At the top of the Bradford third is a thin, hard bed of calcareous sandstone containing a great abundance of marine shells. This layer is the "shell layer" universally recognized by early Bradford drillers. The layer acts as a caprock over the entire third sand. The Bradford third is about fifty feet thick over much of the region, although its actual thickness appears to vary from well log to well log. Most of the Bradford drilling was directed at the Bradford third; perhaps ninety percent of the oil produced in the Bradford region itself were taken from the third sand.

To the north, the productive Richburg sand is very similar to the Bradford third; the two units are probably equivalent. The smaller Crandall sand of Potter County to the east also appears to be equivalent to the Bradford sand. In much of the area outside the major Bradford and Richburg pools, the third is either saturated with salt water or is so tightly cemented that it contains neither oil nor water.

The Bradford Oil Basin

In the north of the Bradford region, the Windfall sand is the next producing unit. To the south, the Lewis Run sand is recognized. These two sands are probably the same, but have never been correlated stratigraphically. These layers are perhaps thirty feet thick. They have been productive mainly in their type-localities at the settlement of Windfall in the northeast and at Lewis Run near Bradford. The two together can be considered as the Bradford fourth sand.

The Bradford fifth, or upper Kane sand consists of sixty feet of thin alternating sandstones and shales. It forms a consistent layer throughout the basin, and is equivalent to the Kane sand elsewhere. The Kane sand of the southern part of the basin, however, is a lower, twenty foot thick sand layer. This unit does not correspond with the Kane sand of the Kane fields, and thus the name Betula Sandstone has been suggested for it. It is coarse-grained, contains clay balls, and is dark brown.

At 2100 feet from the surface of the Chemung is the Bradford sixth or Haskill sand. This is the deepest producing oil sand of the Bradford region, and is defined as the basal member of the Chemung. The Haskill sand is about thirty feet thick; and is most noted toward the south of the basin area.

There are actually many small sand layers in the Chemung group as well. And some of these have been of some importance to oil production in localized areas. Many of them are outliers of the main Bradford basin to the north and west (since the Bradford basin is primarily defined by the extent of the productive third sand). Correlating the various sands stratigraphically is a difficult task, as some are localized, some change form across the basin, and some have several different names even within the same area.

In general, the Warren first and second sands are located above the Bradford first. Between the Bradford first and second lie the Watsonville, Kinzua, Dewdrop

Jon C. Cawley

The Bradford Oil Basin

(Bliss), Cherry Grove (Chipmunk), Tiona sands. And above the Bradford third is the Harrisburg Run sand. Many of these are known more for marker beds in various areas than as productive sands.

At the bottom of the Haskill sand is the traditional division between Chemung rocks and the Portage group below. In Bradford, the Portage group is about 1600 feet thick. Its upper portion consists of dark gray and greenish-gray shales and thin sand layers. The majority of the lower Portage group consists of the Genessee Black Shale member, which is about one hundred feet thick. The shales are part of the dark Hamilton-type shales. They represent a slowly stagnating marine environment which existed before the uplift initiating the Chemung and Catskill deposition.

And below the Genessee shales is the Tulley limestone. This thin limestone layer is much thicker to the south and east, but around Bradford it is only about five feet thick. The Tulley suggests a quiet, almost lagoon-like marine setting within the surrounding anoxic Hamilton shale environment.

The Hamilton group of the Middle Devonian underlies Bradford at a depth of about 4000 feet, and is itself nearly five hundred feet thick. It consists of dark gray and black organic rich shales which are probably the major source rock for the oil in the sand layers above. Oil particles forming in the layers of shale would migrate upward, and would be trapped in the capped layers of sandstone.

The Onondaga limestone layer below makes an excellent marker horizon over the entire Bradford district. It is from seventy to one hundred feet thick, and consists of a very fine, crystalline, light gray to dark brownish-gray cherty limestone. The layer represents a time when the sea was widening over a large, low a Sedimentation was low, and warm conditions probably.

The Bradford Oil Basin

Beneath the Onondaga is the Oriskany formation, a twenty feet thick light gray sand layer which is a major natural gas producer in some localities. The Oriskany consists of sandy river mouth sediments spread across the quiet, limey sea bottom of the Bradford area Lower Devonian.

And the Oriskany is underlain by another sixty feet or so of Helderburg group limestones. These are similar in form to the Onondaga limestones, and are the lowest Devonian rocks in the region.

The Silurian System (443 to 416 m.y.)

The Silurian rocks which lie below the Bradford region are known from well cuttings of deep wells drilled across the region. The upper Silurian rocks are limestones and dolomites associated with arid, isolated Salina Basin environments. And the lower Silurian rocks are shales and sandstones related to the major Silurian rocks of Central Pennsylvania: the Clinton group, the Castenea, and the mighty Tuscarora. The top of the Silurian is 3000 feet deep in the Bradford region.

The Coblestone limestone is a magnesian limestone about twenty feet thick in the Bradford deep wells. It is thought to be the same Coblestone layer as outcrops some seventy miles north in New York State.

The Salina group consists of eight hundred fifty feet of clay-rich limestones and dolomites, which contain beds of anhydrite and rock salt. These beds represent a time when the Pennsylvania sea arm was cut off from the main ocean to the south. Arid conditions prevailed, and the region was turned into an evaporite basin. The salt mines of upstate New York are in these layers of evaporites, which are much closer to the surface further north.

Jon C. Cawley

Mary Margaret
Cawley

Caroline J. (Cawley) Crain

The Lockport dolomite is also associated with this closed basin, and represents an earlier time when conditions were not quite as extreme. It consists of about three hundred feet of very finely crystalline brownish-gray dolomite.

Below the Lockport are about one hundred seventy-five feet of transitional shales known as the Clinton group. These rocks are fine, dark, and calcareous under Bradford. They correspond to the Rochester and Rose Hill of the Central Pennsylvania Clinton group further south.

The underlying Medina group consists of about one hundred thirty feet of red and white layers of hard, shaly quartz sandstone. The upper seventy feet of this group is known as the Grimsby or Red Medina, and is considered to be a northward growth in thickness of the Castenea member above the well-known Tuscarora sandstone of the Central Pennsylvania Ridge and Valley Province. The lower sixty feet of beds are known as the Whirlpool, or White Medina. And these rocks are analogous to the Tuscarora itself, which is the base member of the Pennsylvania Silurian.

The Ordovician System   (505 to 572 m.y.)

The deepest rock cuts within the Bradford basin are the Queenston redbeds of the uppermost Ordovician. In Bradford, the Queenston rocks have been reached at a depth of about 5000 feet. These red shaly sandstones represent early members of the Tuscarora delta and braidplain deposition. They are analogous to the Juniata formation in Central Pennsylvania and are a result of the taconic uplift about six hundred million years ago.

The known stratigraphic column of the Bradford region extends down 5000 feet, and some five hundred million years into the past. Most of the rock layers of interest to oil men are contained in the upper 2000 feet

The Bradford Oil Basin

of the rock mass. Drilling beyond the Bradford third or the Oriskany has been done mostly in an attempt to find other productive deep layers for oil or gas. In more recent years, discoveries of fossil reefs in some deep rock layers in Cattaraugus and Allegany Counties have sparked new searches for gas in the region. And between ten and twenty percent of the gas now produced the Pennsylvania and New York fields comes from these deep well sources.

The history of Bradford has depended almost entirely on its peculiar oil and gas bearing geology. Bradford's past has been built on the fortunes from oil below. And its present and future rest primarily with that flow of oil as well. As the era of secondary recovery draws to a close with the 1980s, population of the Bradford region is decreasing, and Bradford is falling on hard times.

The most optimistic Bradford producers believe that new methods of recovery will come soon to bolster the local production. It is this same belief in oil that has characterized Bradford since its oil beginnings over a hundred years ago.

Jon C. Cawley

"Old Cawley Cemetery" Sligo County, Ireland

Carrowkeel, "Cairn G" Sligo County, Ireland

Jon C. Cawley

Chapter 11
A Brief History of
The Cawley Family

My father, Terence Joseph Cawley always said that the Cawley family emigrated in the 1850s from County Sligo, in Ireland, where they had long been Catholic tenent farmers.

Part of the general legend was that the small place location of "Cawley" in the parish of Ballinafad had been home to the Cawleys since the glacial ice had melted back. My dad always suggested rhetorically that the only thing worth note near to Sligo was a set of Megolithic tombs in the north of the County at Carrowkeel. And that most of the Cawley men *"had been priests long before Saint Patrick ever set foot in Ireland."*

"And that the rest of them had been honest Tavern-keepers."

Our first Peter Cawley had come from Ballinafad, County Sligo, in the diocese of Ail Finn. According to my aunt Rosamond Cawley, this was important, because it was here that Saint Patrick himself had built his first Christian settlement known as Corcoghlan, and the "Tempull Phadruig" in the year 435. The first Abbot bishop of Corcoghlan was St. Assicus, converted from a druidic family, who became St. Patrick's silversmith, casting bells and chalices.

The Bradford Oil Basin

"The features have a boyish or rather girlish look, with a thoughtful expression and very dark eyes; the head not cropped, as perhaps in after-life, but with thick brown and slightly curling hair. The neck is surrounded by the usual laced cuff of the Caroline period of elaborate needlework. He has on a dress of light green, braided with white. Below the waist he wears brown breeches, under an embroidered girdle. There are also traces of a sword belt.

The most observable point in the portrait however, is the hand, with wrist surrounded by an elegantly plaited cuff of lace, of the same pattern as that of the ruff.

Cawley, like Lady Macbeth had a little hand, and like it was imbued with the blood of a king, the difference being that Cawley, like Brutus, helped to deposit one who had been considered a tyrant, to save his country."

from: Cawley, The Regicide.
by: The Rev. Frederick H. Arnold, Ll.B.

The Sussex
Archaeological Collections,
Relating to the
History and Antiquities of the County.

1886

William Cawley, oil portrait, 1620

CAWLEY'S ALMSHOUSE, CHICHESTER.

Jon C. Cawley

My dad would occasionally point out the somewhat apocrophal story of the British scions of John Cawley, a brewer, and the mayor of Sussex in the early 1600s. John and Mary Cawley's eldest son was William, a member of Parliament, and later one of the 59 who signed the Death Warrent of King Charles the First in 1649.

My father used to shrug and say: "A lot of people have some nobility or aristocracy in their family line... But there are a lot fewer who have had a rightful King of England put to Death."

Quoting J. T. Peacey, William Cawley, Oxford DNB, 2004:

"In 1649, William Cawley was appointed to the High Court of Justice for the trial of King Charles. He attended every meeting of the court and signed the King's death warrant. During the Commonwealth, he became very active on parliamentary committees and was zealous in buying up the estates of former Royalists.... but he opposed the extreme religious sects... ...He fled abroad at the Restoration in 1660 and finally joined Edmund Ludlow at Vevey in Switzerland, where he died in January 1667."

Accounts say that Cawley was duly buried at Vevey and the antiquarian records his epitath as:

"Hic jacet
Tabernaculum terrestrial
GULIELMI CAWLEY
Armigeri Anglicani
Nup de Cicestria
In Comitatu Sussexia
Qui postguam aetate
Sua inservivit
Dei consilio
Obdormivit
6 Jan 1666
AEtatis suae 63"

The Bradford Oil Basin

Later reports however state that the regicide Cawley was eventually exhumed, his body encased in lead to preserve it, and that it was eventually returned to the Cawley family Almshouse in Chichester by his sons:

"No. 8 Cawley's Vault: In the account of the opening of this vault in 1882 SAC Vol XXXIV page 21 note I alluded to a previous examination of it in 1816 as remembered by an aged inmate This has been recently verified In looking over some old Court Books of the Chichester Workhouse the Master Mr Stratton came on an entry confirmatory of it as being of interest in several respects is here subjoined:

On Wednesday the 24th day of April 1816 a vault was accidentally discovered in repairing the pavement of the chapel of the Poor House and was opened in the presence of Mr Joseph Smith Mr Edward Gilbert and Mr Wm WickhamJun who were the Guardians of the Poor The inside contains the fragments of two wooden coffins and the remains of their inmates also one made of dipped lead the shape of the human body in length 5ft 10in 1ft 10in over the breast and eight inches at the feet containing a perfect skeleton supposed to be that of William Cawley the regicide who signed the warrant for the execution of King Charles the First and was the founder of the Chapel and Almshouse.

Although history informs us that at the restoration of King Charles the Second he was obliged to fly to Holland where he died in poverty still it is very probable that he might in his last moments express an anxious desire to be buried in the place of his nativity and in the sepulcher, founded by himself and his corpse might have been smuggled over for that purpose."
Sussex Archaeological Collections, Volume 36.

Jon C. Cawley

The Younger William, was born in 1628, studied law at Oxford, and eventually wrote a comprehensive book on British law as it pertained to Catholicism in Ireland, which was published in 1680. Its complete title was:

"The laws of Q. Elizabeth, K. James, and K. Charles the First concerning Jesuites, seminary priests, recusants, &c., and concerning the oaths of supremacy and allegiance, explained by divers judgments and resolutions of the reverend judges : together with other observations upon the same laws : to which is added the Statute XXV Car. II. cap. 2 for preventing dangers which may happen from popish recusants : and an alphabetical table to the whole, by William Cawley of the Inner Temple, Esq."

It is said that this book, while hailed as a surprisingly complete work, indeed, also ironically enabled the Irish Catholics for the first time, to defend themselves effectively in the English courts.

In later years, a cousin of my dad's cousins by the name of Sharon Anne Dalrymple became very interested in topics geneological regarding the Cawley family history. Dalyrymple Made a pilgrimmage to Sligo to search for Cawley family tombstones and records from the parish.

From Sharon Dalyrymple's newsletter, circa 1986:

"I have been doing extensive research on the Peter Cawley/ Cauley family in Pennsylvania. It looks like the name was originally Cawley in Ireland but sometimes changed to Cauley in the U.S. Most of my current research is in Pennsylvania.

Peter Cawley came from Co. Sligo, Ireland around 1850. Born around 1819, he had at least 2 brothers, John and Terrance. He had a sister, Ann, who married Patrick Bly (Blygh) in PA.

He also had 7 half brothers who settled around Boston, Mass. Peter married Catherine McKenzie (Kensey) from Westport, Co. Mayo, Ireland, and had 14 children, 5 who died as children at Sartwell, Pennsylvania. Of the remaining 9, four because priests at St. Patrick's in Erie, PA, two became nuns, one sister was their housekeeper for 70 years at St. Pat's.

The remaining two, John (my great-grandfather) and Mary married. Other related families are Crowley, Plunkett, Burgoyne, Hanley, Bly and Welch. Would love to make contact with anyone knowing this family. I have spent 2500 hours researching them myself and I have written a newsletter and also a book (on the priests)."

Sharon and my dad compared notes and photographs over several seasons in the 1990s, and Sharon Dalrymple published her Masters Project on the Cawley family, and in Particular the Cauley priests of Erie in 1993. Much of the information in this manuscript is derived from both my dad's notes, my Aunt Rosamond's notes and stories, and Sharon Dalyrymples manuscript from 1993.

The Irish Settlement
Sartwell, Rock Run, and Turtlepoint

Until the late 1700s, the side valleys north of the Allegheny river, above the great periglacial "Buffalo Swamps" and "Shawmut" were hunting lands of the Seneca, while the lands south of the river were held by the Susquehannock. Along the Allegheny ran the Forbidden path, demarking the region's tribal migration route.
By 1796, the pacts and treaties with the natives had been broken by George Washington, who in turn encouraged new settlement of the lands north of the Allegheny river.

Jon C. Cawley

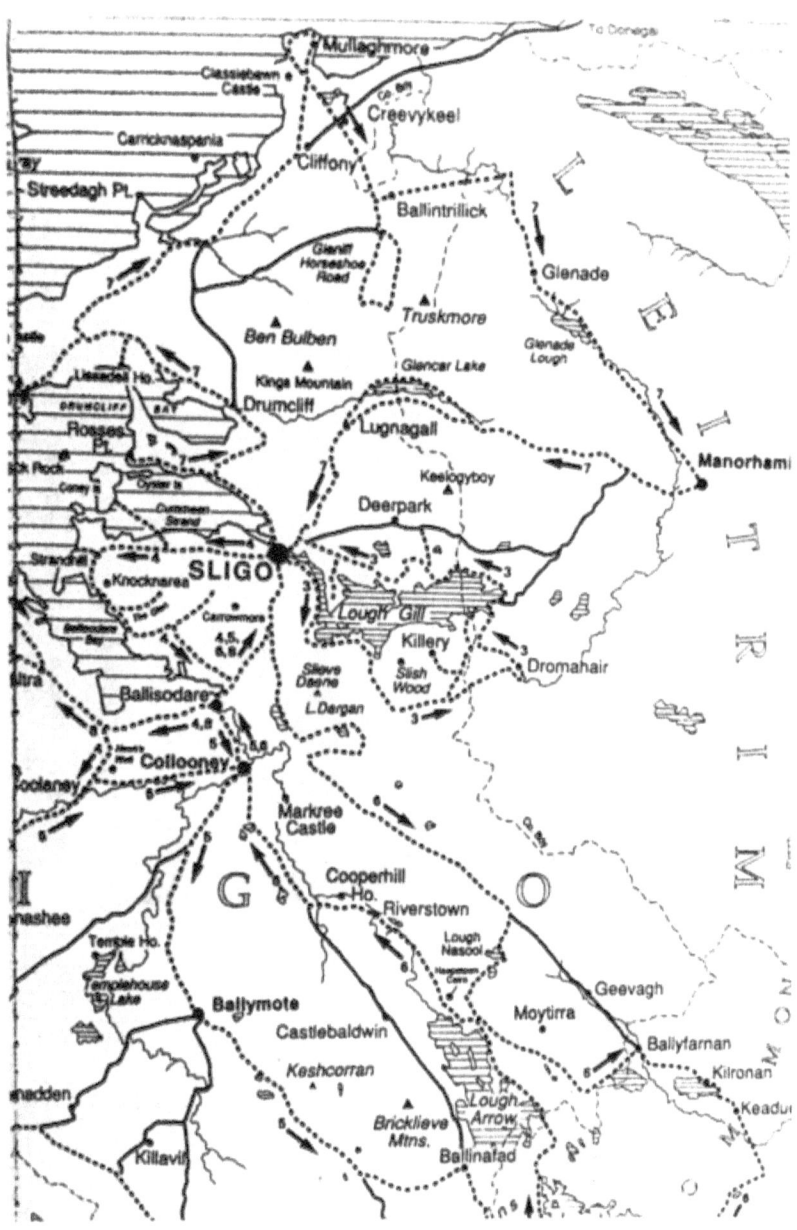

Sligo, Ireland map, showing Sligo Town (middle), ruins at Carrowkeel, (top), and Ballinafad (bottom). Map after that shown in Sharon Dalrymple, 1993.

The Bradford Oil Basin

Early view of the Irish Settlememt at Turtlepoint, from a post card.

The Bradford Oil Basin

The Cauley Priest Family. (around 1894-1897)

Standing:
Father Joseph; Sister Bernardine; Father Peter; Sister Bernadette; John J. Cauley; Father Stephen.

Seated:
Father Charles; Cassie Cauley; Catherine)(McKenzie) Cauley; Peter Cauley; Mary (Crowley); Terrence Cauley.

Jon C. Cawley

John Keating had been born in Ireland in 1760. In 1796 he and his family moved to the United States, he purchased 297,000 acres of land in McKean, Potter, Cameron, Clinton, and Clearfield counties. In 1842, through the efforts of Keating & Company, seven Irish families settled in the wilderness of Sartwell. The families were George Oliver, James Oliver, Edward McMann, Patrick Masterson, John Crowley, William Crowley and John Connors.

They cleared the land for the cemetery and had built a rudimentary log church. From this time on, the holdings at Sartwell, Rock Run and Turtlepoint were known as The Irish Settlement.

By 1856 several families, including the Cawleys, Peter (2) and Catherine McKenzie, had moved to the settlement and subcribed funds to the project.

Peter Cawley and Winnifred Kelly

Our progenitor Peter (1) Cawley was born in County Sligo in Ireland and spent his early life there. He married Winnifred Kelly, and in Ireland, had five children: Peter, Ann, Terrance, John H., and Catherine. Most or all of these eventually immigrated to the United States, and settled either in New York State, or at the Irish Settlement at Sartwell, Rock Run and Turtle Point.

Later, after the death of Winnifred (Kelly) Cawley, he remarried Winifred Keaney, and had a second family of children, Dominick, Timothy, Tom, Patrick, Mike, and Jim.

At midlife, Peter (1) emigrated from Ireland and apparently followed the construction of the canal system in upstate New York. According to Rosamond Cawley, he entered at Boston, meeting up with one or more of his siblings who had arrived before him, and had settled in that area. He originally worked at quarrying stone in upstate New York, around Rochester.

The Bradford Oil Basin

Rosamond said that although he was of limited literacy, Peter (1) was skilled in stone cutting, and that he had had showed interest in collecting birds' nests, plants, and fossils from the quarry stones.

Of the Cawley children, Peter (2) Cauley was to be the father of the several Cauley priests and nuns. While John H. Cawley was to begin the lineage that leads to myself (Jon Clayton Cawley).

Peter Cauley and Catherine McKenzie

Peter (2) Cauley first married a Mrs Welch, a widow with four children. We assume that this marriage occurred in Ireland. From this first marriage, Petr (2) had a son, whom they named Terrance (2) Cauley. It appears that Mrs. (Welch) Cauley died soon thereafter.

Peter (2) Cauley (1819-1903) arrived in the United states around 1850 and at the age of 31. It is not clear that he necessarily traveled with the elder Cawley, or that the two spent specific time together. Still, the fact that several of the Cawley siblings made the trip within a short time period may suggest some family coordination.

Peter (2) Cauley now in the States, married Catherine McKenzie in 1852 at Portage, New York. Catherine was quite a bit younger than Peter. Immigration records suggest that Catherine had come from Europe in November of 1846, "Departed from LIverpool on the Denvonshire" and arriving at New York City. She is listed as having been about 12 years old at the time, and as having traveled with several of her siblings. We are told that Catherine's family was extremely religious, with this McKenzie line having produced many priests and nuns in Ireland

Their eldest son, they also named Peter (3), technically the third, after the paternal grandfather, from Sligo, Ireland. He was born at Rochester, Dec 1856.

Catherine (McKenzie) Cauley
Born July 25, 1833, Westport, County Mayo, Ireland
Died May 27, 1916.

Trinity
Churchyard,
Erie,
Pennsylvania

Jon C. Cawley

By 1858, Peter (2) had cleared and settled a farm property at Rock Run, and moved the family there soon after, first seasonally, and then permanently. Eventually, with nine living children, they settled down to a life of farming at Rock Run and Turtlepoint.

Peter (2) and Catherine's Children were:
1. Cauley, Peter M. 1854-1938 New York
2. Cauley, Mary (Crowley) 1856-1911, Rock Run
3. Cauley, Winfred (sis Bernadetta) 1856, Rock Run
4. Cauley, John, J. Aboutt 1858, New York
5. Cauley, Joseph M, b. Sep 1862, Pennsylvania
6. Cauley, James, b. Abt 1865,
7. Cauley, Stephan H, b. Nov 1866-1954 Pa
8. Cauley, Charles, b. Abt 1869, Pennsylvania
9. Cauley, Rosamond (sis Bernardine) 1854
10. Cauley, Frank, bur. Sartwell, Pa.
11. Cauley, Ellie, bur. Sartwell, Pa.
12. Cauley, Catherine, bur. Sartwell, Pa.
13. Cauley, Catherine A. (Cassie) 1872-1963
14. Cauley, Terrance (born of Mrs. Welch)

The eldest son, Peter (3) Cauley counselled and decided to enter the Catholic priesthood, encourged, says Sharon Dalyrymple, by Catherine, who was a bit of a zealot. With time, the majority of the living siblings followed him to Erie, and the Parish church of Saint Patrick. And in 1894, Peter and Catherine turned their lands at Rock Run over to their son John J. Cawley, and also moved to Saint Patrick's in Erie. In the mean time, oil had been discovered in the region, and so resources had begun to come to the family.

Once at Erie, all reports are that the older Peter (2) lived a quiet life. Records indicate that for some 10 years, the only title Peter (2) would accept at the parish was that of "Gardener"--and that he enjoyed being the official bell-ringer of the church.

The Bradford Oil Basin

Peter had been young during the famine years in Ireland, and he apparently could not bear to think of anyone ever going hungry. To that end, one of the mottos of the Irish parish in Erie during the Cauley years was "that no-one should ever go away hungry or unfed."

The elder Peter (2) Cauley died in 1903, and is buried beneath a small, plain stone at Trinity Cemetery at Erie, near his wife Catherine, and near to the lake. The obituary of Peter (2) Cauley reads:

"An Exemplary Christian Gentleman has gone from Among us. Aged Father of St. Patrick's beloved pastor is called away from beloved family circles. At the ripe old age of 84 Peter Cawley sr. passed away, last evening, at the home of his sons, the Reverends Peter, Joseph and Stephen Cauley, where he had been an honored and cherished resident for nearly nine years.

Peter Cauley was born in County Sligo, Ireland, early in the century which has just closed, and came to this country when he was about 22 years of age. Here he entered business as a contractor on the New York State Canal. He afterwards bouth land at Turtlepoint, McKean County, Pa. and devoted himself to farming an employment to which he devoted himself actively until taking up his residence in Erie in 1894.

Nr. Cauley was, with the exception of one severe attack of illness about 18 years ago always in the best of health until the stroke came which precipitated the illness which has just terminated fatally. After this former acute illness, his health had been perfectly restored and through the past three score and ten years allowed him to maintain, when he became a citizen of Erie, his vigor and energy which were pleasant indeed to remark as he came and went between the church where so much of his time was spent and the pastoral residence.

He was an example and an edification to all as a constant and devoted attemdant at church services, and

Jon C. Cawley

The Bradford Oil Basin

and as the faithful and energetic caretaker as to the ringing of the bell, the beautification of the church lawn, tasks in which he seemed to delight as in theire way related to God's service.

Although a most kindly and courteous gentleman, he seemed to feel that the closing years of his life, freed from the exactions of business, were given him as a time of retreat and meditation, and though he made few acquaintances here, he nevertheless, because of his beautiful example, leaves many sincere mourners.

He suffered a second paralytic stroke on Friday last, June 26, and had been since that time perfectly helpless, but his death had not been expected until within the last two days.

There survive: His wife and children as follows: Rev. Father Peter M. Cauley; twin sisters, one of whom is a Franciscan sister, known in religious life as sister Bernadette now in Buffalo. and Mrs. Daniel Crowley of Turtlepoint. Amother daughterm also a Franciscan religious sister, Sister Bernardine: John Cauley of Turtlepoint; Revs Joseph and Stephen Cauley, Saint Patrick's Church; And Miss Cassie Cauley."

The Cauley Priests:
Monsignoir Peter M. Cauley

It is said that the Roman Catholic Church was sprung from the line of "Peter the Rock." So too, the spiritual and Catholic lineage was risen from the role model of Monsignior Peter (3) Cauley, long associated with Saint Patrick's Irish Parish in Erie Pennsylvania.

Perhaps the most complete short description of the Cauley family priests is to be found in Nelson's biographical dictionary and historical reference book of Erie County, edited by in 1840:

The Bradford Oil Basin

"The Reverend Peter M. Cauley, pastor of Saint Patrick's Roman Catholic Cathedral parish, Erie Pa., was born in Rochester New York, December 18, 1855 and is a son of Peter and Catherine (McKensey) Cauley. His parents are natives of Ireland but came to this country before marriage, which took place in Portage NY: they now reside with Father Cauley. Mr Cauley followed the business of quarry man, but in 1858 he removed to McKean county Pennsylvania and engaged in farming, where he remained until 1894 when he came to Erie. The family consists of ten children:

Terry (foreman in the lumber woods for Col Kane of McKean county Pennsylvania); Rev. Peter M.; Winnefred and Mary A. (twins), the former now Sister Bernardetta of St Francis Asylum Buffalo NY, and the latter the wife of Mr Daniel Crowley of Turtle Point Pa and the mother of nine children; Rosa, Sister Bernardine, also of St Francis Asylum; John who is engaged in farming at Turtle Point; Rev. Joseph M., who was educated at St Bonaventure's College, ordained December 25, 1893 and is now assistant at St Patrick's parish; Stephen, a student in St Bonaventure's College; Charles, a student in the Erie Business University; and Cassie, a pupil in Villa Marie.

Father Cauley received his early education in the public schools of McKean county, Pennsylvania and his philosophy and theology in St Bonaventure's College, where he matriculated in 1879, and from which he was graduated in 1887.

He was ordained in St Patrick's Cathedral, Erie uly 24, 1887. He was then assistant at Titusville Pa., and Warren Pa., successively remaining in each place three months. He was next located at Oil City, where he remained one year. After a few weeks passed at Sartwell Pa., he was placed in charge of a mission at Conneautville Pa., where he remained four and one half years. He then passed ten weeks at East Brady, Pa., after which he came to Erie in 1894. It is evident that Father

Jon C. Cawley

Monsignoir Peter M. Cauley
December 18, 1855 - November 20, 1938

The Bradford Oil Basin

Cauley not only comes of a reliious family, but his work indicates that he has marvelous organizing talent that finds scope in his latest field of labor."

Father Joseph H. Cauley

From Sharon Dalrymple's Masters Thesis: The Cauleys Answer God's Call, 1993:

"Father Joseph H. Cauley was the first family member to join Father Peter at St. Pat's. He was born September 23, 1863 in Turtlepoint, McKean County, Pennsylvania. He came to St. Pat's one month after Father Peter just as soon as he was ordained a priest. He had the distinction of being the first priest ordained at new St Peter's Cathedral in Erie.

He served St Pat's for 26 years until he died suddenly in 1919. During that time he was given an assignment in charge of St. Francis church for 16 months. But he returned to St. Pat's. He was instrumental in getting the Auditorium for St. Pat's built on time. With Father Peter becoming so ill during the ocnstruction of the new church at St. Pat's. Father Joseph was invaluable in helping to overcome obstacles to its completion. On December 17, 1918 there was a double celebration. That day celebrated the 25 years of Father Joseph as a priest and 25 years of Father Peter as pastor of St Pat's....

...Shock hit the city of Erie when Father joseph died in 1919. He had been in good health just three days earlier when he was at a meeting in the mayor's office making plans for the homecoming of the Eightieth Division. He developed cerebral menengitis and pneumonia after a monor surgery. Although he seemd to have been at Death's door and was even given Last Rites, he made a rally and was on his way to recovery when Death struck. In 1922, Father Peter converted a Baptist church into a community service center.

Reverend Joseph H. Cauley
September 23, 1863 - May 23, 1919

Green alabaster Holy water fount having belonged to Father Joseph. (This is likely the one given to him by the Franciscans at Saint Bonaventure during his time there.)

Jon C. Cawley

It was dedicated as a memorial to Father Joseph and was named the Josephinium."

Aunt Rosamond always said that it was ever Father Joseph who was expected to inherit the main mantle of Saint Patrick's if and when Father Peter stepped aside. He was a philosopher, He was brilliant, and he of all the brothers had his finger on the pulse of the civic community. From Rosamond's stories, we always had the impression that Father Joseph had died tragically young. In fact, Father Joseph was 56 years old when he died. Young, surely, but not tragically so. I think that it was as much the shock of how quickly he died that made it such a shocking event. Everyone's assumptions about the ongoing role of the very vital and alive, energetic Father Joseph were spectacularly shattered within a week.

The Auditorium of Joseph Cauley served the Community for many, many years. eventually the building and the site was purchased by the Erie Insurance Company. There is a related story about the role of the Cauµey's in the Erie Insurance company, about how when the civic minded need for honest insurance was an issue in the city, it was the Cauley priests who loaned the money to start the company, which remained loyal to St. Pat's and the Cauleys through all the years.

After purchasing the old Auditorium, Erie had the building inspected, and it was deemed unsafe for further use and occupation. The building was razed, and a new greenspace park dedicated to the Cauley priests was planted on the location.

Reverend Monsignoir Stephen H. Cauley

According to Sharon Dalrymple:

"The Right Reverend Monsignoir Stephen Cauley was the next brother to join in the fight for survival of St. Pat's.

The Bradford Oil Basin

Born November 11, 1866, at Trutlepoint, McKean County, Pennsylvania, Father Stephen came to St. Pat's upon his ordination on May second, 1897. He had enrolled at Bonaventure College in 1890 and completed his studies there in 1897..."

Father Stephen was the scholar and academic of the group. Being younger, Stephen spent much of his time playing the supporting role as the assistant early on. But with the eventual demise of Peter and of Joseph, eventually he was to reach the primary role of power at St. Patrick's. In the mean time, he spent a good portion of his energies supporting the church school and the mission of the orphanage. He designed curriculum, and acted as Chaplain and principle.

"...One of Father Stephen's main roles at St. Pat's seemed to be as caretaker and janitor of the facilities. A normal day for him would start when he would stoke the fires and turn the gas heat on in the basement of the church at 5:15 a.m. before offering Mass at 6:00 each week-day at St Joseph's Orphan Asylum. He would end the day by locking up the church (and later the Auditorium) each night at 11:00 p.m. In 1898, he fell twenty-two feet when he went to fix the broken rope of the church bell tower. It was rotten and old and broke. Relatives of a parishioner who had just died complained that the bell had not been rung for him. That night, after the fall, Father Stephen was back at his work hearing confessions."

Dalrymple also points out:

"Another duty that Father Stephen took on was the dreation and publication of the parish annuals. These gave parish reports, activities and pictures showing what went on during the year. He started this in 1899, and it was continued until 1972. There is a wealth of information on the church and its history in these annuals."

Reverend Monsignoir Stephen H. Cauley
November 11 1866 - May 17, 1954

Father Charles Cauley
August 9th, 1868 - 12 December 1950

Jon C. Cawley

"When Father Stephen died on May 17, 1954, an era came to an end. He was the last surviving brother of the Cauley family. He had been ill for four months. He was the oldest priest in the diocese at the age of 87."

Father Charles Cauley

"Father Charles Cauley was the last of the brother priests to join Father Peter at St. Pat's. He was born August 9th 1868 at Turtlepoint, McKean County Pennsylvania. He attended Catholic High School in Philadelphia. Father Charles, unlike his brothers, took a detour before becoming a priest. He went to Erie Business University. He 'had successfully established himself in business as a pioneer in the oil industry before he answered the call of the Divine Master and, therefore, had no need to seek security in the priesthood.' He was ordained as a priest in 1905 at the age of 37. He immediately was assigned as an assistant to St. Pat's.

Father Charles had different interests and duties than his brothers. In the early years he took an active interest in the athletics for the young. He developed one city championship baseball team and many basketball champions. He managed girls basketball teams which won five city championships. He developed many other sports activities. He was involved with various church organizations eg. heading the Holy Names Society.

Over a period of 42 years, heard 336,260 confessions. That is an average of 22 confessions for every single day of those 42 years 'Bishop Gannon identified Father Charles' priestly life as primarily associated with the confessional box, with his hand almost constantly for 45 years raised in absolution, his kindly heart pouring forth confort, and his wise counselling sending penitents away better and stronger. Even in an hour of tragedy he was capable of sending a man away with a smile on his face through his little touches of humor'...

The Bradford Oil Basin

...In 1945, he developed Pheumonia from La Grippe. He was given no hope for survival but survive he did. After a three month stay in the hospital, there was a sudden change. He recovered and continued to work for St. Pat's for another five years until he died in 1950."

Sisters Bernadine and Bernardette Cauley

There were two sisters both of whom took religious orders: Rosamond, twin of Mary, who was later to become Sister Bernadine of the Sisters of Saint Francis. And Winifred. She also entered service with the Sisters of Saint Francis, and became Sister Bernardette.

In 1989, Sharon Dalrymple wrote to the Superior of the Sisters, at Williamsville, New York, to enquire about the Cauley Sisters. The following is excerped from the letter of reply:

"I was able to find Obituary notices of both Sisters which seem to tell a lot. Sister Bernadine (Rosamond) entered our Order on August 2, 1882; received habit 2/22/83 and made her vows on March 19, 1885. She gives her mother's name as Cath Kensy and born in County MAyo, Ireland. She was Superior at St. Agnes, St. Al's, Springville, and also at Holy Family Home many years. In fact I think she was Superior at the time of her death at Sisters' Hospital on March 22, 1934. (this was before my time in the convent.)

Sister Bernadette (Winifred) was born in Lockport and entered our Order October 2, 1880, received habit 8/2/81 and made vows on August 2, 1883. She died at Saing MAry of the A. on August 3, 1945. She gives her parents - Peter Cawley born in County Sligo, Ireland, Mother, Cath McKensie, Born in Westport Ireland. (I dont know which is correct, her information or that of Sister Bernadine.) Enclosed is the copy of her obituary too.)

Jon C. Cawley

The Bradford Oil Basin

The Cauley Rectory Building.

Jon C. Cawley

Miss Cassie Cauley

Finally, at Erie was Catherine, or "Cassie" Cawley According to Sharon Dalrymple:

"One final member of the Cauley family served at St. Patrick's. Yet she is hardly mentioned anywhere, although her contributions were just as important as the rest. Cassie (Catherine) Cauley was one of the sisters of the priests. She chose to give up her life to serving St. Patricl's and her brothers by becoming their housekeeper and bookkeeper. She never married. Between her and her lifelong friend Mary McGinty, they managed to keep the rectory running smoothly. They both lived in the rectory, and were inseparable. They died one month apart.

The both worked very hard and deserve credit that seems to have been denied them. The "Pictoria Review of Fifty Years" which was written to commemorate Father Stephen's fifty years with St. Patrick's does not even mention her name, except for two short sentences, even though she gave more years to St. Patrick's than any of the other Cauleys. She was the last survivor of that generation.

She joined her brother, Father Peter, shortly after he became pastor of St. Pat's in 1893. She served as housekeeper for 70 years, until her death in 1963. She may not have been in the limelight like her brothers, but the work she did made it easier for her brothers to do theirs.

From Saint Patrick's church bulletin, July 26, 2015:

"One hundred years ago this week, the four Cauley brothers and their sister, Cassie, moved into the newly completed St. Patrick's rectory. Cassie would serve as the housekeeper until her death in the 1960s. Built at a cost of $57,000 the house was largely financed through the generosity of the Cauley brothers as well as their father. The parish contributed only $12,000 of the final cost. The house has 34 rooms, 11 bathrooms, 8 fireplaces, an elevator & functioning steam room.

The original house had 20 Waterford crystal chandeliers, all but three of which disappeared in the 1970s. The second and third floors originally had eight two-room suites for each priest as well as a bathroom for each suite. Up through the 1990s, the house was largely full of priests but is now mostly empty. Offices fill the first floor and the third floor is an apartment for the pastor."

John J. Cauley

According to Sharon Dalrymple:

"It is said that John J. Cauley (Sharon's great grandfather), the only son who married, was the one designated to carry on the family name, so he did not become a priest. He was born December 1860 in Turtlepoint, McKean County, Pennsylvania. He married Mary Honora Welch in 1887. They had ten children, one who became a nun (Sister Bernard). He went into the oil business and bacame very affluent from it. Eventually he moved to California and lived next door to Joan Crawford. Tradition says that he would place a Saint Joseph's medal on the ground where he would place oil tanks for a new oil well. The only time that he did not strike oil was when he did not put the St. Joseph's medal there."

"According to an interview I had with Bill Cauley, first cousin to my mother, John Cauley was a true Christian in every sense of the word. He had strong values and ethics. He went to chruch every Sunday. He did not swear or drink but he was not adverse to a fight when he deemed it appropriate. The oil business was a rough business and in California your employees needed a healthy respect for your fighting abilities if they were to cooperate. John would immediately fire any employee he caught stealing or drinking. He believed in honesty and integrity, even when it hurt.

John was generous. When a man was killed working for him on his many oil rigs, John, out of the goodness of his heart, would see that the widow and children were financially taken care of for as long as they needed it. Even after John moved to Hollywood, his house was always open to anyone, including his children's friends. John and his friends and family would gather around the piano and sing. John enjoyed a good pipe of tobacco.

John J. Cauley, Mary (Welch) Cauley and their brood.

Saint Mary's Roman Catholic Church, Sartwell, Annin Township, McKean County, Pennsylvania

Jon C. Cawley

Chapter 12
Oil and Destiny:
The Cawley Brothers

On the other side of the Sartwell Cawleys, there were always a few opinions that maybe the Cauley priest family might have been trying to buy their way into heaven on the strength and wealth of Bradford oil.

After all, old first generation Peter (1) Cawley did have other children besides Peter of the priestly line.

The siblings, born in Ireland to Old Peter and Winnifred Kelley, were: Peter (2) Cauley (1819-1903); Ann Cawley (Bligh) (c. 1820-1892); Terrence (1) Cawley (1823-1864); John H. Cawley (1827-1910); and Catherine Cawley (c. 1829-?)

Terrence Cawley

Peter (1) Cawley's second son was named Terrence (1), who had been born in Ireland, and had come to America somewhat earlier than his father. That Terrence had settled at Sartwell eventually as well. The mother of Terrence (1) was Winnifred Kelly.

My father, Terence (3) Joseph, was fond of relating a tale about old Terrence. It is said that about the time that the Sartwell Saint Mary's Congregation was being set up, a contingent of Cawleys (probably led by the new wife, Catherine) went up the valley to pay Ter-

Saint Mary's Roman Catholic Church,
Sartwell, Pennsylvania--Interior

Jon C. Cawley

rence (1) a visit. It seems that he was not attending Mass regularly, and not tithing regularly to the Church.

Terrence Cawley is reported to have said, *"So long as I could see this valley every morning at sunrise, I'd not ask Saint Peter to use his keys for me!"*

Apparently he was almost excommunicated! My father, who had been an altar boy, had stepped away from the Church in adult years, and he always somewhat relished the nonconformist bent of his namesake ancestor.

The son, Terrence (2), was born in 1848, and died in 1902. It is this Terrence who is listed in several sources as having served as a lumbering foreman for General Kane. He was somewhat of a loner, avoided the oil boom, and is listed as having located in Olean for some years of his life. There is no indication that either of these Terrences ever married.

John H. Cawley and Mary Carberry

John H. Cawley was born in Ireland to Winnifred Kelly. He emigrated from Ireland about 1849 and settled at Rock Run, Pennsylvania. His wife-to-be, Mary Carberry was born in ireland 1837. She emigrated to America about 1846 with her mother and three brothers. She married John H. Cawley, and they had eight children.

Children of John H. and Mary (Carberry) Cawley:

1. Sarah Cawley 1857-1938. Married Joe Finn, and had seven children.
2. James Cawley 1859-1936. Married Nellie Appleby and had 10 children.
3. Peter (4) Cawley 1861
4. William J. Cawley 1863-1947. Married Lin Diehl and had four children.
5. Mary (Mame) Cawley 1869 - 1941. Married Martin Joseph Welch and had four children.

Jon C. Cawley

Mary (Mame) Cawley Welch, and Martin Joseph Welch

6. Teresa Cawley 1872.
7. Matilda (Till) Cawley 1875-1960 Married Jack Hevenor and had three children.
8. Ambrose Vincent Cawley 1878. Married Lucy Bruder and had 1 child.

An obituary from January 27, 1941 issue of Bradford Evening Star:

"Mrs. Mary Welch, 71, Dies at home Sunday after an extended illness. Eldred, January 27. Funeral services for Mrs Mary Welch who died at 9:30 a.m. yesterday will be held at 9:00 a.m. Wednesday at St. Raphael's church. Burial will be in Sartwell cemetery. Death was ascribed to complications.

Mrs. Welch was born March 14, 1869 at Turtle Point. On January 14. 1889 she was united in marriage to Martin J. Welch, who died about a year ago. She was a member of St. Raphael's church, and had lived in Eldred for 19 years.

Surviving relatives include three daughters, Mrs. William Rosenswie of Eldred; Mrs. W. P. Mulvey of Buffalo and Mrs. Dudley Hull, of Allegany, a son, J. P. Welch of Eldred, 14 grandchildren, two great grandchildren: a sister, Mrs. John Hevenor of Eldred two brothers, William J. Cawley of Turtle Point and Ambrose Cawley of Port Allegany."

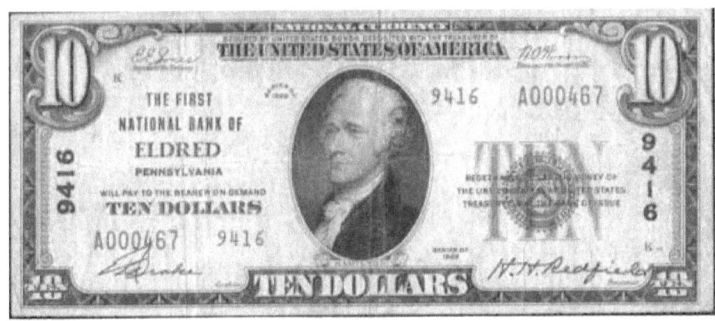

The Bradford Oil Basin

James Cawley and Nellie Appleby:

Aunt Rosamond would say that Perhaps her father, James and her mother Nellie may not have been quite in love with the oil industry as the Cawley children turned out to be. She always said that James had seen the effects of the Standard on the individual producer in the Bradford field, and he was wary of outside interests. James always liked the ideal of being something of a gentleman farmer and a businessman.

When oil was discovered on the Cawley farm properties at Rock Run and Turtlepoint, he was glad to accept the windfall to help raise his family. But he soon sold the oil patch itself, and purchased another farm out beyond the edges of the field at Bolivar New York. When the Bolivar field was subsequently opened, and the Cawley land at Bolivar was found to have oil under it, James and Nellie eventually moved back to Eldred.

That having been said: is worth considering that the following small notes from the Bolivar Breeze newspaper:

"February 26, 1909: J. H. Cawley of Bolivar has sold his farm of 165 acres at Turtle Point to his brother, A. V. Cawley, but he has reserved the oil and gas rights."

"September 25, 1919: SPECIAL COLUMN:. FOR SALE: J. H. Cawley farm, one mile southwest of Bolivar, together with oil and gas rights and good timber. Total of 210-acres, about 100 acres under cultivation. Royalty on this property amounts to a good sum, many, locations-to "flood." For price, etc., inquire of F. J. McDermott, Bolivar."

"February 02, 1928: J. H. Cawley, formerly of Bolivar, has been elected a director of the First National Bank of Eldred, Pa."

Jon C. Cawley

The Cawley stone at St. Mary's Cemetery, Bolivar, New York.

Jon C. Cawley

Rosamond told the story that once James and Nellie had made an extended trip to New York City. While they were away, the several sons, including my grandfather Clayton, had set up a nitroglycerine operation in the back yard of the Bolivar house. When James returned, he was awfully mad: he informed the sons that if they wanted to get blown to pieces, that was their own fool business-- but that if they were going to do that, they needed to at least have the respect to do their bathtub nitroglycerine in the next valley over, and not in his back yard where it was going to endanger their mother, and the family home!

According to Rosamond, the Cawley brothers provided nitro and shooting services for several years around Bolivar. Apparently they ran a clean operation, with very few remarkable incidents.

Common and somewhat apocryphal family legend had it that they once lost one nitroglycerine load and wagon on top of the hill to Olean. The horse and wagon were found blown to pieces; the driver, meanwhile was never located.

Rosamond said that the following investigation showed that the man had fought seriously with his wife in the weeks and months before the incident. and the conclusion was that he may have stood well back and blown up the load, and left for other parts unknown. Rosamond always pointed out, however, that the Cawley Brothers agreed to pay the funeral and insurance benefits to the widow in any case.

Finally, in 1936, the Bolivar Breeze newspaper carried the formal obituary of James H. Cawley:

May 7, 1936: James H. Cawley, 77, of Eldred, a former resident of Bolivar for 14 years died Saturday at his home after a short illness. Funeral services were held at St Rachel's. church at Eldrd at 10 o'clock Saturday morning, with the Rev. Father Liebel officiating.

Burial was at the Catholic cemetery at Barden Brook, the pallbearers being nephews of the deceased.

Mr. Cawley was born March 19, 1859, at Rock Run, Pa., the son of John and Mary Carberry Cawley. From the time of his marriage in 1887 to Miss Nellie Appleby, until 1905 he was a resident of Sartwell, Pa. He came to Bolivar in 1905, where he had oil interests, and then to Eldred, where he had since resided.

During his residence in Eldred he was a member of the Eldred borough council and a director of the Eldred First National bank.

He is survived by his Wife; three daughters, Mrs. F. R. McDermott and Mrs. C.V. Ebert of Bolivar, and Miss Rosemary (sic) Cawley of Eldred; six sons, Mark; Homer, Marvin, Clayton, Randall and Howard Cawley, all of Eldred.

Twelve grandchildren also survive: Among those from Bolivar attending the funeral fete: Mr. and Mrs. C Ebert and daughter, Frank McDermott and daughter, Mary, Mr. and Mrs. Francis. H. McDermott and daughter; Mrs. Frank Appleby...: Morris Appleby, Mr. and Mrs. Joseph Hughes, Mr. and Mrs. William Hughes, Mr. and Mrs. Sheldon Appleby..."

Nellie (Appleby) Cawley was known to be the stalwart and stoic wife of James. The two cut a handsome couple whether at Bolivar, Eldred, or in New York City. James was thin and stylish, and Nellie is described as being level-headed and thoughtful. She was strong enough to see through raising her brood of Six sons and three daughters. Photos of Nellie always show a serious and not always a happy person. We know that she had health difficulties, and a note in the Eldred Eagle in March 1927 states:

James and Nellie (Appleby) Cawley, left;
Martin and Mame (Cawley) Welch on right.

The Bradford Oil Basin

"Mr. and Mrs. Mark Cawley and Mr. and Mrs. J. H. Cawley motored to Buffalo Thursday where Mrs. J. H. Cawley consulted a specialist. She has been in poor health for the past several months."

In fact, my great grandmother Nellie lived to the ripe old age of 85, having been born in 1870, and having died in 1955. Aunt Rosamond, who stayed at home and took care of Nellie at Eldred often talked about Nellie's depression in later years. According to Rosamond, all of the Appleby's had beautiful teeth, and Nellie was no exception. She cared for them with baking soda and salt. In her very last years, she would frequently obsess about whatever would happen to her if she ever lost her teeth; how could she ever possibly get along. Rosamond pointed out that When Nellie died at 85, she still had her own teeth. Nellie is buried at the Saint Mary's Cemetery at Bolivar.

The Applebys:

Nellie Appleby was the younger daughter of Sheldon S. Appleby (1843-1918) and Mary Etta (Hall) Appleby (1846-1922). Sheldon had served in the Civil war, and had seen action in Virgina and other locations up and down the Eastern seaboard. Eventually Sheldon S. and Mary Etta had 12 children, 7 of whom lived to adulthood.

Sheldon Appleby's response to a questionaire for his Civil War pension in the National Archives, Washington, DC, dated 4 July 1898:

"I am married. Wife's maiden name Mary Ett. Hall." When, where, and by whom were you married? "Sept. 2nd 1865 at Friendship, N Y by Rev. C. P. Clark." Names and dates of birth of his living children: "Emma A. Appleby Born June 17th 1867 Nellie C. B. Dec 9th 187_ Frank L. May 12th 1872 Charls [sic] R Sept 29th 1877 Lewis M. Aug. 6th 1879 Samuel R., Jan 9th 1888, Jay, June 19th 1889"

Mary Ette (Hall) Appleby and
Sheldon S. Appleby

The Appleby Homestead House at Bolivar, New York. Sheldon S. and Mary Ette on porch.

In the 1900 census, the Appleby's were based in Bolivar New York. The following information is given: Emma Adda (32) is listed as a cook in a hotel; Nellie C. (30) listed as "Mother of children"; Frank L is not listed; Charles Randall is listed as an oil well worker; Lewis W. is listed as an oil well worker; Samuel Stanton is listed as being at At School; George Jay is listed as being At School.

The older daughter "Aunt Emma", a dedicated spinster and sufferagette, was always described by Rosamond as having been the one who "had saved her life" when Rosamond was a teenager. Aunt Emma painted fine porcelain plates, which was all the rage at the time; she had apparently studied to do so with the elder Mrs. Rhoades, who (later) kept the Rhoades variety store in Eldred.

Above: Mark and Mary (Hughes) Cawley
Below: Howard and Coletta (Butler) Cawley with Elaine.

Jon C. Cawley

Above: Robert Cawley, son of Mark amd Mary (Hughes) Cawley.

Below: Olive Cawley with John F. Kennedy.

Terence Joseph Cawley and Grandmother Nellie (Appleby) Cawley.

James H. Cawley son of Jim Cawley, grandson of Mark and Mary (Hughes) Cawley.

Jon C. Cawley

The Children of James and Nellie (Appleby) Cawley:

James and Nellie had ten children:

1. Mary Leona Cawley 1888-1985. Married Francis McDermott and had one daughter, Francis Ellen.

2 Mark Leo Cawley 1890-1970. Married Mary Hughes and had six children; Jim, Margaret, Bob, Joan, Mary Alice and Bernard.

3. Homer F. Cawley 1892-1976. Married Nellie Gardner and had two children; June and Genevieve.

4. Eva Cawley 1894-1968. Married Cyril Ebril amd had four children; Vincent, Bernadine, Rita and Barbara.

5. Marvin Cawley 1897-1958. Married Mary Turock.

6. John Cawley 1899. "Lived for three days"

7. Clayton J. Cawley 1900-1946. Married Margaret Long and had one son, Terence Joseph.

8. Rosamond J Cawley 1903-1998.

9. Randall J. Cawley 1905-1971. Married Nora A. Butler and had one daughter; Katherine Ann.

10. Howard Vincent Cawley 1907-1959. Married Coletta Butler, and had one daughter, Elaine.

The Bradford Oil Basin

Rosamond J. Cawley

Rosamond Cawley was effectively the Cawley family matriarch for over fifty years. Born in Bolivar, and moving with James and Nellie and the rest fo the family back to Eldred in about 1920, Rosamond was the defacto care-giver in raising the large family. She lived most of her later adult life at the Cawley homestead house on Mechanic Street of Eldred. And she took care of her mother Nellie in declining years into the 1950s.

By trade, Rosamond spent her life as a teacher in the Otto-Eldred school system. She earned her standard certificate for elementary education from Lock Haven State Teachers' College. She began her teaching career at the small one room schoolhouse at Sartwell, near to the Cawley homestead lands, and near to Saint Mary's Sartwell church. And then taught fifth grade in the Eldred system for the next 42 years.

The Bradford newspaper listed its faculty for the 1933 school year as follows:

The Bradford Era, August 28, 1933; Eldred. Aug. 27 Eldred High school will resume on September I following the regular summer recess with the township school opening for the 1933-34 term on Sept. 11. Harold Childs is principal of the high school and Paul Bundy of the township school. The school faculties: High school French and science. Mary Jones. Salamanca; commercial and English, Sylvia Lyons, Bradford; mathematics and science. Forrest Cummings, Eldred; Latin and English. Richard Shattuck, Smethport; music, Miss Davis. Grades First, Mary Cantwell. Eldred; Second.

Dorotha Clark. Eldred; Third. Norma Butler. Eldred; Fourth, Phyllis Enright. Bradford; Fifth. Rosamond Cawley. Eldred; Sixth. Ruth Carpenter. Eldred; Seventh. Cecil Gamble. Port Allegany; Eighth. Angeline Marrone, Eldred.

The Olean Times Herald, Wednesday September 4th, 1940. Pqge 4: Faculties At Eldred Pennsylvania are Listed: Eldred Borough School returned to work with one new teacher. The faculty for the term follows: Supervising Princlpal, Harold T. Childs; Assistant Principal and Supervisor of music, Miss Wildamary McInroy. of Jersey Shore.

English and Latin, Miss Ann Kubllewicz, Forest City. Science, Joseph West, Mansfield. Commercial, John Macsuga, Ashley, Pa. Mathematics and English, Gino Beldonl, Eldred. Eighth Grade, Miss Lettie Austin, Mansfield. Seventh Grade, Cecil Gamble. Sixth Grade, Miss Ruth Carpenter, Larabee.

Fifth Grade, Miss Roamand Cawley. Eldred. Fourth Grade, Miss Alice Drake, Shinglehouse. Third Grade, Miss Norma Butler. Eldred. Second Grade, Miss lona Black, Haymaker. First Grade, Miss Dorothea Clark. Eldred. Substitute teachers High School, Mrs. Ruth Stoughton; grades, Mrs. Frances Lynch.

The Township Consolidated School staff is as follows: Supervising Principal and Eighth Grade teacher, Paul L. Bundy. Seventh Grade, Miss Mary Welch. Sixth Grade, Mark Scarcell. Fifth Grade, Miss Zena Gulnac. Third Grade, Miss Claire Paul. Second Grade, Mrs. Marian Foster. First Grade. Miss Harriet Kymer. Dental hygienist, Miss Frances Wood. Teacher for extra pupils from the grades, Miss Emma Marks.

The Bradford Oil Basin

Another decade and the War had well ended. The Eldred Joint School system roster begins to look somewhat more like it would for many, many years. Literally, from the era of unpaved streets and horse-drawn vehicles to the era of the American moon landing, Rosamond and her colleagues made up the priesthood of teachers who taught and civilized two or three generations of Eldred citizens.

The Bradford Era, August 29, 1949; Eldred Borough, Township Schools To Reopen Sept. 6. Announcement has been made that the Eldred Borough school and the Eldred Township school would reopen for the 1949-50 term on Tuesday, Sept 6. faculty members for the schools were also announced. Attention was called to the new state law passed by the last session of the legislature that all beginners must be at least five years and seven months old as of Sept. 6. The following teachers will comprise the faculty for the Eldred Borough School: Harold T. Childs English and Latin, Mrs. Ann Breck Science and Mathematics. Mrs Emaline Childs Mathematics, French and History, Joseph Wolcott. Music, Mrs. Wildamary Leffler First Grade, Mrs. Mary Shields Second Grade. Mrs. Claire Marts Third Qrade Mrs. Norma Oeuder Fourth Orade Miss Ruth Carpenter Fifth Grade, Miss Rosamond Cawley Sixth and Seventh Grade. Mrs Angeline Petrusie. Mrs. Wildamary Leffler, music teacher will also teach English in the grades.

The Eldred township school faculty follows: First Grade, Mrs Bernice Irons Second Grade, Mrs. Minnie Reese Third Grade. Mrs. Jeanette Loop Fourth Grade. Mrs. Louise Dennis Fifth Grade, Miss Zena Gulnac Bixth Grade. Mark Scarcell Seventh Grade, Mrs. June Fowler Eighth Grade, Mrs. Mary Welch. Extra Teacher, Mrs Hasel Robson.

The Bradford Oil Basin

The Bradford Era, December 10, 1952, Eldred: 56 Persons Help Eldred Resident Observe Birthday: Fifty-six members of the Cawley family gathered at the home of Nellie Cawley at Mechanic Street to help her celebrate the 82nd anniversary of her birth. Dinner was served at 8 p.m. from a table centered with a large personalized birthday cake. The Cawley family resided at Rock Run for a while and later at Bolivar, New York, before moving here in 1920. Cawley was born in 1870 and She was married to the late James Cawley at Marys Catholic Sartwell in 1887. Among the guests were eight 10 grandchildren and 11 great grandchildren.

Through all of the years, Rosamond was a staunch role model, teacher of culture and knowlege, of proper decorum. She was a vampire hunter. She improved penmanship; she graded english essays on citizenship, and future dreams, and fifth grade windows on the world. The good students could look up to Rosamond as pointing them to the hidden knowlege of the library, or to the proper person to talk to about a topic or a problem. Poor students were terrified, since the fury of Rosamond the teacher could also know no bounds. This was tough love indeed.

Bradford Era (Newspaper) - December 16, 1965, Bradford, Pennsylvania. Elementary schools of Eldred Borough, Eldred Township, Duke Center and Rixford will be humming with activity today and Friday with the presentations of Christmas programs. The Eldred Township's program will be at 7:30 p m. today while Eldred Borough's will be Friday, starting at the same hour. Duke Center's program is at 1:30 pm today and Rixford's at 1:30 p m

Miss Rosamond Cawley, left, who is retiring after 42 years of teaching, and Mrs. Norma Geuder, who is retiring after 41 years. Gifts were presented to the honored guests by Larry Miller, president of the PTO. (Era Photo by Stewart)

Friday. Director of all of the presentations will be Mrs. Wildamarv Leffler, elementary music supervisor for the Otto Eldred Joint School system. Assisting are teachers of the various grades.

Faculty members of the Eldred schools include Mrs. Harold Hansen, Mrs Mary Shields, Miss Ruth Kennemuth Mrs Norma Gender, Mrs Angeline Petruzzi, Miss Rosamond Cawley, and Edward E. Harrington. Assisting at Duke Center are Mrs. Alta Woodmancy, Mrs John Jeannette, Mrs Leonard Sanbergm, Mrs. Ebert Lundin, Mrs. Maude Kroner and Norman Kelly. From Rixford Mrs .Eilean Holcomb, Mrs. Wardloe Wilcox, Mrs Anson, Mrs Esther Nelson and Mrs. Gladys Rainor.

The Bradford (Pa.) Era, Monday, May 13, 1968 page 5. Miss Rosamond Cawley and Mrs. Norma Geuder. retiring elementary teachers in the Otto Eldred Joint Schools were honored Sunday at an open house held at the elementary school. The event was sponsored by the Eldred Boro Parent Teacher Organization.

Miss Cawley was graduated from Eldred Boro High School, Class of 1922. She attended Lock Haven Teachers' College and obtained a permanent certificate in elementary education.

She taught her first year at Sartwell in Eldred township, and the remaining 41 years in Eldred Boro Elementary School as a teacher in the fourth and fifth grades.

Nrs. Norma Gueder was graduated from Eldred High School, class of 1926. She attended Clarion and Mansfield State Teachers' Colleges and obtained a permanent Standard Certificate in elementary education.

She taught her first two years in Moody Hollow School in Eldred Township, and the remaining 39 years in Eldred Boro third grade.

Gifts were presented to the honored guests by Larry Miller, PTO president, who also had charge of the guest book. Red-tipped carnation corsages were presented the teachers by Mrs. John Welch.

Mrs. Welch and Mrs. James Sullivan were co-chairmen of the reception, assisted by Mesdames Orton Maynard, George Brown and Ronald Gueder, and Miss Ruth Kenemuth, publicity.

Jon C. Cawley

School Grounds, Penna. R.R. & Allegany River Valley, Eldred, Pa.

Window Glass Factories, Eldred, Pa. (1907)   Pub. by J. W. Grassfield, Eldred.

Jon C. Cawley

The Bradford Oil Basin

Gusher in Tulsa Oil Field, Tulsa, Okla.

Jon C. Cawley

Clayton J. Cawley

Rosamond related that in 1922, a young Clayton, along with a couple of the other brothers took a new open top car, purchased from Berg and Todd of Eldred, and made a long trip across country to Oklahoma, lending their oil field expertise to the opening of a new oil field discovery.

They set up their business, copied directly from the Cawley Brothers pattern in McKean County, and bringing with them several specific inventions and innovations from the Bradford field. It was rural and empty land, and everything from steam engines to drilling casing had to be trucked in from outside distance.

As Rosamond relates it, during one of the drilling projects, one of Clayton's very good friends managed somehow to get his hand and arm trapped in the travelling band belt from the steam power to the bull wheel of the derrick. The force of the moving belt was quick and inexorable, and separated the man's arm from his torso. He was alive, bleeding out, and in terrible pain.

Clayton and the brothers knew basic first aid, but not so much for grevious wounds of this sort. They quickly prepared a red hot iron from the forge and attempted to cauterize the site of the missing arm. The man went into shock immediately, and died in Clayton's arms.

According to Rosamond, Clayton was quiet and thoughtful for awhile after the accident, and then announced that the others could do what they would, but that he was getting back into that car and heading back to McKean County. All of the brothers agreed, and so they returned home to Pennsylvania. Rosamond said that Clayton never really got over that death of his friend and colleague, and that it very much of haunted him.

The Bradford Oil Basin

In the Oklahoma oil field, 1922.

Jon C. Cawley

The Bradford Oil Basin

Clayton was in some ways the most adventuresome of the Cawley boys. He was trained as an oilfields geologist (and in the building of this project, I ran across several of his notebooks, full of well logs, still on file at the Smethport Courthouse). Clayton was fascinated by the explosives side of the trade, and seemed to have less fear of the ntroglycerine than the others. He was careful and thoughtful.

He was also an inventor. In our family we still have various patent papers and photographs of Clayton with his cylindrical flying machine--as well as of his cam-based "non-tangle-able dog chain."

The year my grandfather turned 27, he was logging a well near eldred, when a chained iron bailer broke loose from up inside the derrick tower. It pivotted down and struck Clayton on the head, crushing a portion of his skull. He was rushed unconcious to the Mountain Clinic in olean (much of a days travel) once it was determined that he wasnt going to die immediately.

The Bradford Oil Basin

Jon C. Cawley

It was quite a mystery, and Clayton lay insensible and in a coma for seventeen days before waking up. At that time, old Dr. Mountain told him that he had better make peace with his Creator, and to do the things in life that he might be waiting for-- because at most he might have twenty years to live. But that he also might just fall down dead at some point along the way.

In the seasons that followed, Clayton proposed to my Grandmother, Margaret (Long) Cawley--Soon they had birthed my father in Movember 1936.

Clayton was something of an invalid in following years, and experienced recurring migraine headaches. He was well provided for by the Cawley Brothers company, which continued. And Clayton primarily played the role ofconsultant and businessman, rather than doing much field work at all. I was told by my grandmother Cawley that he did a fair amount of writing, and that he still kept his rock and fossil collection.

In 1942, Eldred was struck by a massive summer flood, which filled Main Street to nearly a building story deep with water and mud. Clayton manned one of the rafts to help make sure that people were safe, and could get at least portions of their belongings from the raging Allegheny River.

Meanwhile, as events led up to the Second World War, The Catholic Priest of Saint Raphaels and Saint Mary's gave or sold his Philco floor model short wave radio to my grandfather. Since Clayton was physically somewhat limited by then, he served as the Civil Defense Radio monitor. He helped arrange for town blackout drills and air raid drills.

The War ended, but then in the autumn of 1946, my grandfather was stricken with an anurysm, and he died on the last day of September, 1946. My Dad was 9 years old.

The Bradford Oil Basin

Friday October 04, 1946   Eldred Eagle
Clayton J. Cawley passed away Tuesday morning.
Death came suddenly after a brief illness. Funeral yesterday

Word Tuesday morning of the death of Clayton J.Cawley, 46, came as a shock to his host of friends in this community where he had spent the greater part of his life.

Mr. Cawley was taken suddenly ill Monday night at about 9:30 O'clock at his home on South Main Street and was immediately taken to the Mountain Clinic, Olean, where he died at about five o'clock Tuesday morning without regaining contagiousness. A cerebral hemorrhage was believed to have been the cause of his death and also believed to be the result of a head injury received about nineteen years ago.

While not enjoying the best of health for some time he appeared at all times to be in the best of spirits and by his jovial nature his relatives and friends were not fully aware of his failing health.

He was born May 20, 1900, at Turtle Point, son of the late J.H. Cawley and Mrs. Nellie Cawley. Several years ago he moved to Eldred with his parents from Bolivar N.Y.

He is survived by his wife, Margaret Long Cawley, his mother Mrs Nellie Cawley, one son Terrance Joseph Cawley, Eldred; three sisters, Mrs L.McDernott, Mrs Cyril Eberl. Bolivar; Miss Rosamond Cawley, Eldred; five brothers, Mark Cawley, Homer Cawley, Randall Cawley and Howard Cawley, Eldred; Marvin Cawley, Olean.

The sympathy of a host of friends is extended to bereaved family in their great loss.

The funeral was held yesterday morning at eight-thirty from the home and at nine o'clock at Saint

Jon C. Cawley

Raphael's Church , with Father J.H. Davis officiating at the Requiem Mass. Interment was made in the Catholic cemetery at Bolivar N.Y.

The funeral was largely attended by relatives and friends.

Pall bearers were Harry Frohnapple, Bryan Lynch, Burt Murphy, Elmer Petruzzi, Cleary Slavin and Roy Strickland.

The Bradford Oil Basin

Draw Knife and Barrel Awl belonging to Andrew Gutzler and later, Arthur Long.

Arthur H. Long and Louise Marie (Gutzler) Long.

Louise Marie Gutzler Long

Arthur H. Long, my Great Grandfather, was born in America on July 6, 1883-- As an adult he was an oil fields cooper, making barrels and wagon wheels in the Bradford basin. It was said that he also had an early stake in the Prouty baseball bat factory in Eldred. He was a tall silver-haired man of 68 at the time of his death in 1951.

His wife, Louise Marie Gutzler Long was also born in 1883, and she died in 1975 at the age of 92. Her mother was Cecelia Marie Belocker Gutzler who was born in France in 1844, and died at Warren, Pennsylvania in 1922. The father of Louise was Andrew Gutzler. also listed as a cooper and barrel-maker. I still have his iron spoke-shave with which he did woodworking, and his hand-forged barrel awl, which my great grandmother, and then my grandmother had.

The Gutzler family was from Alsace, at Bas Rhin, and we were always told that the Gutzlers traveled together as a family from the old world to the new. The story was always told on holidays that somewhere in easternmost France, the extended family had just set down to dinner, and were about to say the evening prayer, when a bullet was shot from outside. It came in through the roof rafters, and then dropped straight down onto the plate of the family patriarch.

There was a long silence-- then the Father said: "After dinner, begin to pack your things-- we are going to America."

They emigrated through France, but as they were leaving the harbor, it was announced that War had broken out between France and Germany. The pilot of the ship was just beginning to leave the harbor, and he kept on the journey instead of turning around as ordered. And so the family came to the United States.

Above:
Eleanor, Louise, Molly (lap), Abe, Chuck, Grandma Cecelia Gutzler, Francis.

Below:
Molly Long Baby photo.

Jon C. Cawley

It was always told to us that the Marie middle name was actually after a progenitor Marie Madeline Gutzler, who had been born in Bas Rhin in 1770.

I remember "Grandma Long" as a thin, gaunt and patient elderly woman. I remember her at the time of the moon landing in 1969, musing on what it was like to watch men land on the moon, when in the same lifetime she remembered the Wright Brothers' airplane launch at Kitty-Hawk in 1903! In later years, illness had paralyzed my great grandmother's legs, and she used an old-style wheel chair to get around her house. She would very seldom ever go any farther than the end of the little sidewalk to her house. It was a prim little white wooden house on Edson Street, surrounded with many tall ferns, and with a huge arching apple tree in the back yard.

"Olean Times Herald: Mrs. Arthur Long, Eldred, PA. Mrs. Louise Long, 92 of Edson Street died Friday August 8. 1975 in the Bradford Hospital following a long illness.

She was born July 30, 1883 at Warren, a daughter of Andrew and Cecelia Marie Belocker Gutzler. She married Arthur Long who died in 1951.

Mrs. Long was a member of Saint Raphael Catholic Church of Eldred and was an honorary member of its Altar Rosary Society.

Surviving are three sons, Bernard A. Long, Charles A. Long, and Gerald R. Long, all of Eldred; five daughters, Mrs. Margaret (Molly) Cawley of Eldred, Mrs. Frank (Eleanor) Wilkinson of Custer City, Mrs Howard (Catherine) Crowley of Lewis Run, Mrs Gerald (Francis) Brown of Horseheads, New York, and Mrs Joseph (Geraldine) Perone of Greenville LI; and 22 grandchildren, 36 great grandchildren, and a great-great grandchild.

Friends may call at the Frame Funeral Home, Eldred, after 7 pm today. A Mass of Christian Burial will be celebrated Monday August 11. 1975 at 11 am in Saint Raphael Church with the Rev. John M. Fischer, pastor as celebrant."

Katherine,
Eleanor,
Louise,
Abe,
Bob, Molly,
Francis.
c.1959

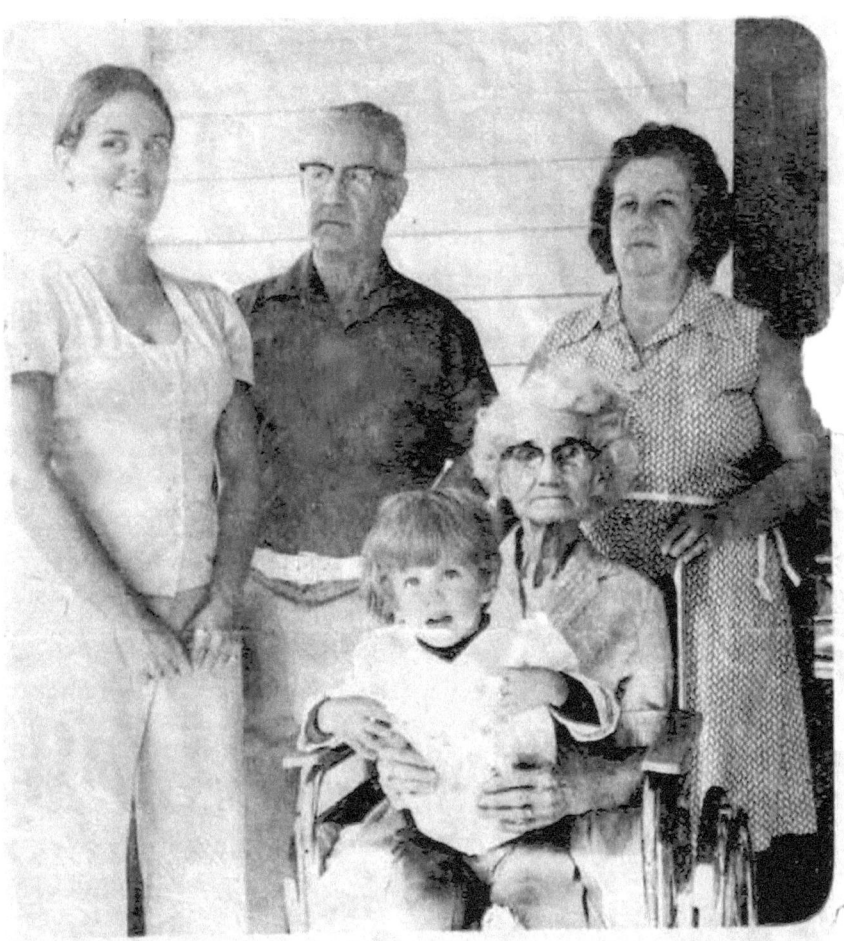

Eldred Woman Celebrates 90th Birthday

FIVE GENERATIONS of the Long family were caught by the Times Herald camera at Eldred this past Sunday. The big occasion was a family party in celebration of "Grandma" Long's (Mrs. Louise Long of 56 Edson St.) 90th birthday celebration. Among the 40 family members at the affair, held at Mrs. Long's home, were from left: Mrs. Richard (Diana) Mayer of Arcade, N.Y., great-granddaughter; a son, Charles Long of Eldred; Mrs. Francis (Jean) Copeland of Arcade, granddaughter; two-year-old Jason Mayer of Arcade, great-great-grandson seated on the lap of Mrs. Long. Friends and relatives attended from Eldred, Arcade, Lockport, Long Island, Lewis Run, Custer City, and Olean.

(Times Herald Photo)

The Bradford Oil Basin

Margaret L. Cawley

So, my grandmother was named Margaret Louise, with her middle name being her mom's. She was one of eight Children. First there was Eleanor, and then Bernard (Abe, 1905-1990), and then Charles (Chuck, 1907-1990) and then Catherine, next Francis and then Margaret (Molly) and then later on, there were Gerald (Bob) and Geraldine (Sally), who were twins.

I always think of my grandmother Cawley as standing by Cuba lake gazing off into the distance. Or maybe busying herself making chicken and biscuits in a cast iron pan in her small kitchen. I think of her reading, as she often spent her evenings. Or sewing, concentrating, on her sewing machine in the front bedroom.

My Grandmother always went by the nickname of Molly in Eldred. I dont think that she ever cared for Margaret. She was handsome, resembling the best traits of both of her parents. She had piercing dark eyes and dark hair. While she was growing up, Eldred was in its oil renaissance, and she was one of the somewhat privileged town girls. She could walk to the town high school, and just about anywhere within the few square blocks that made up Eldred-town.

She was always quick and proud to say that she and her girlfriends had made a point to see every single Hollywood motion picture that had ever come to the Eldred Theater! The silver screen was definitely magical-- and represented dreams and culture of faraway and exotic places Outside.
The girls all knew and kept track of the comings and goings of all of the stars and starlets.

After All, Louise Brooks, the silent film star was from nearby Rochester, New York (and Ingrid Bergman also lived there in the 1940s). Grace Kelly was originally from Philadelphia; while Catherine Hepburn was from Williamsport, Pennsylvania, just down the road a ways.

Kerosene lamps having belonged to my Great-great Grandmother Cecelia Marie Gutzler; Later belonging to my Great Grandmother, Louise Marie Long. Collection of the Author.

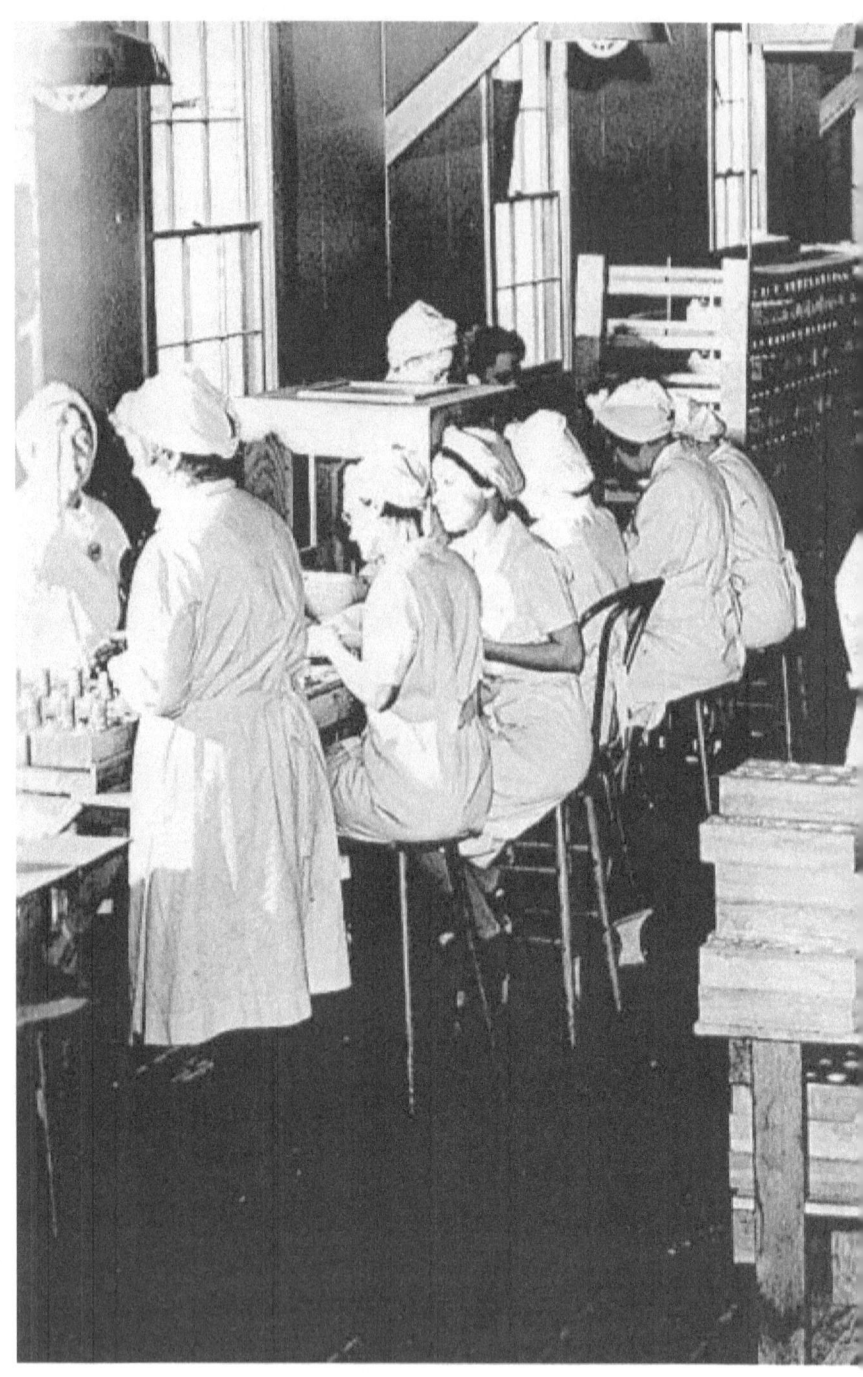

Women of Eldred working at the Munitions Plant.

Jon C. Cawley

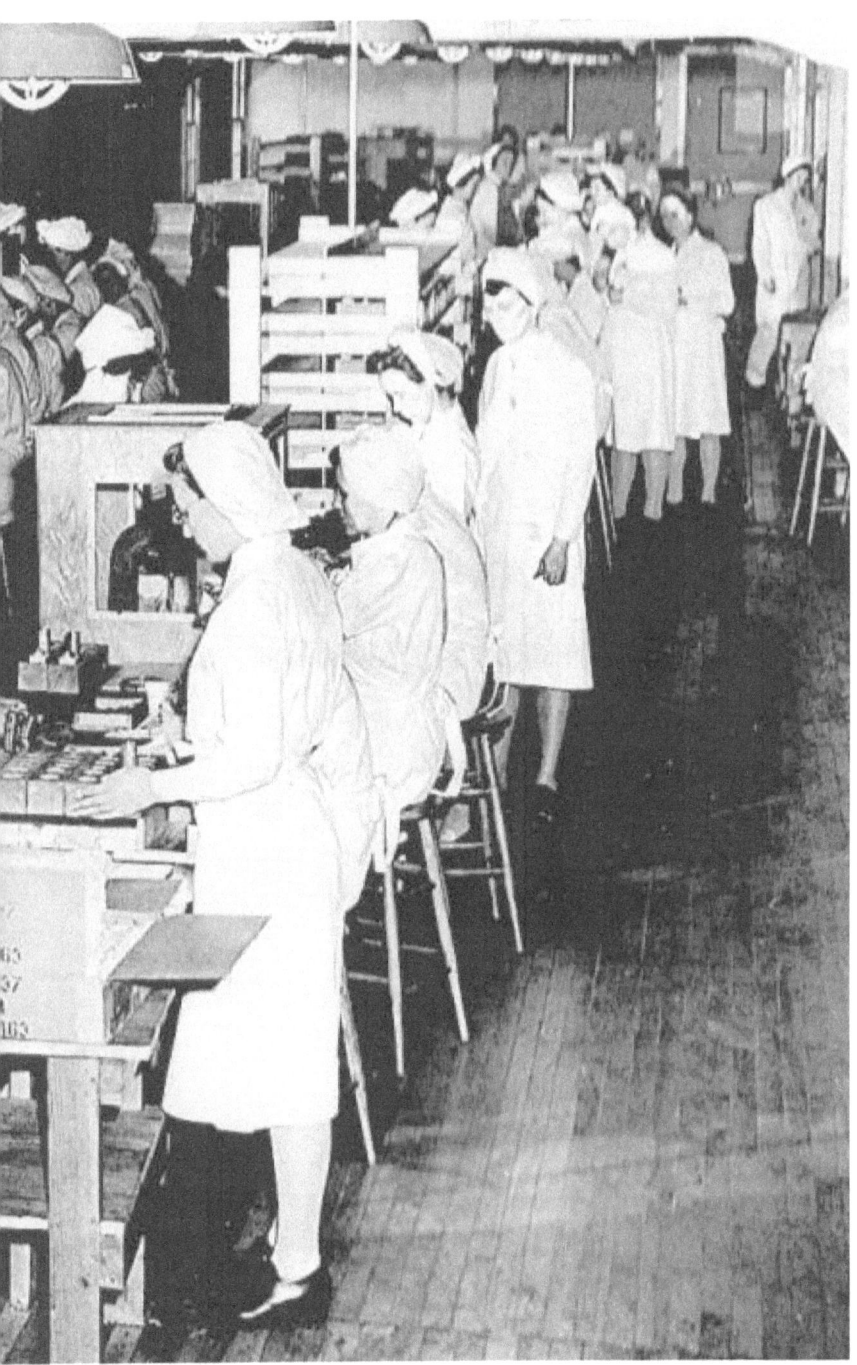

Courtesy, Eldred Historical Society

Jon C. Cawley

If my grandmother had such aspirations early on, as many did-- she kept them to herself. She was pragmatic and practical, taking life as she saw it. She married Clayton J. Cawley, ten years her elder. They purchased the Phalen House on Main Street for cash. They set up housekeeping, and soon had a child, my father, in early November of 1936.

Despite the Great Depression, the Cawley Brothers were busy keeping up with oil well production in the Eldred valley. And so there was money for small luxuries, even though my grandfather's earlier head injury limited how much physical labor he coukd actually do. The Phalen house had come already mostly furnished, with its oriental rugs and overstuffed Victorian furniture from a generation before. The Cawleys settled in.

Friday nights would often entail a trip by car or by train to nearby Olean or Bradford or Ohio for a special dinner, or a fish fry, or for an occasional dance.

In the summer of 1942, the huge summer flood washed through Eldred, and the Phalen House was thoroughly flooded. My grandmother afterwards dutifully dried out furniture and carpetings and sorted through water-logged photographs, saving what she could. It was a tough haul, but they survived it.

With the War Effort, my Grandmother began to work at the National Powder factory on the edge of town.

In 1946, however, my Grandfather Clayton died from the anurysm caused by his head injury. This left Molly as a young widow with a child. In retrospect, there are indications that my grandparents had known that it was coming. That my grandfather's health had been declinig quickly.

He had doted on my Dad, buying him toys and books to replace those which had been lost early by the flood.

The Bradford Oil Basin

He bought my Dad a Charley McCarthy ventriliquist doll, and most of a set of the "Bomba The Jungle Boy" series by Roy Rockwood. He bought a full set of "The Books of Knowledge" and a fine Encyclopedia Brittanica that my Dad wouldnt be able to read for several years. Clayton took his son fishing as often as possible, teaching him about the trout stream, and the moods of the Allegheny River. Each season they took in "fresh-air" children from the City, to keep my father company.

After Clayton's death, my pragmatic grandmother rolled up her sleeves and went back to work at the National Powder dynamite factory below Eldred. There she became a front-office secretary and switchboard operator for the explosives works.

My mom, Caroline Crain relates in her **"Place at the Table"** memoires, a story about the big explosion at National Powder in 1968:

"As I have said, just north of Eldred eas located the National Powder Company where a good amount of dynamite was produced to supply the needs of a growing nation. There are tales of several explosions through the years, and of men being blown to bits and maimed and blinded. One such explosion happened while my mother-in-law Margaret (Molly) Cawley worked in the front office. She had worked ther for many years without a major mishap. The office was located in a separate building near the highway."

"At the first yells of "She's gonna blow!" one summer morning, those three office ladies wasted no time diving under their big oak desks, and covering their heads, prepared for the worst. After an explosion that blew most of the leaves off the surrounding trees, and shattered most of the windows, they eventually got the call that all was clear. That time no-one had been seriously injured. Still, everyone was summarily dismissed for the day. Molly was not a drinking person, but she said that she went directly home and then poured herself "a little something" to calm herself down. After a brief call to let us know that she was all right (as news of such an event travels fast in a small town), she went straight to bed."

Molly sold the big old Phalen house, downsized and bought a much smaller cottage, built in 1915, at 32 Edson Street. My Great-Aunts Rosamond and Coletta helped out both physically and financially, especially taking care of my Dad during times when Molly needed to work.

As a young widow, Molly was stylish in the mode of the times. She was independent; she kept her own finances, kept her own car, ordered things from catalogues and magazines, clipped coupons and recipes. She sewed her own clothes. She embraced all the rituals of family; she kept the holidays, and went to Mass regularly at St. Raphaoel's. She never remarried.

She cooked, and within the family, Sunday Dinner was an expectation. She had a Hoosier cabinet full of baking supplies, a Siamese cat named Timmy, houseplants, roses, and a weeping willow tree.

Through the years Molly bought new furniture and remodelled the kitchen and bathroom of her little bungalow. She traveled to Olean by car, and sometimes to Erie or Buffalo by train. She kept her sound social standing in town, patronizing Mrs. Farris' ladies clothing store on Main Street, and Miss Rhoades' Variety Store across the street. She bought her groceries every week at the A&P. She always banked with the security of the First National Bank of Eldred, and she always bought her cars at the Todd Garage in Eldred.

Molly believed in Eldred. Believed in the passing small town society. On Memorial Day, we would watch the parades, Watch Uncle Abe and the other Great War veterans march. Listen to the lonesome notes of the Bugle, which were echoed back by the hidden bugler on a hill-

The Bradford Oil Basin

side far away. Every Memorial Day Molly put flowers on all of the graves, and spent a little quiet time alone with Clayton's grave at Bolivar.

When my father married and had children, Molly stepped in as a willing Au pair. She traveled with us on trips and vacations to help take care of us kids. She was proud to show us off, and she always took time to visit all of the various elderly relatives about town, so that we would know and remember them.

Jimmy Cawley, who was the Postmaster; Rosamond who was the Schoolteacher; Uncle Homer, who lived by the Church; Uncle Randall, who smoked a cherry-wood pipe. Abe and Jeanette Long; Chuck and Ruby Long; Uncle Bob and Aunt Kay, who were stylish, and had white carpeting and a ceramic bull-fighter and bull on top of their new RCA color television set.

Molly lived her whole entire life in Eldred. She understood it, and she understood that it was somehow frail and needed tending against the rush of time. As our grandmother, she went to all of our school plays and assemblies. She lived and died in her own home, and on her own terms, in 1986.

The Bradford Oil Basin

Terence (3) Joseph Cawley

My father, as a young man, roamed the mountains and meadows of McKean County, hunting grouse and woodcock of the young woodlands and abandoned farms of the post oil boom. He had a coon-hound named Major whom he loved. He had a plethora of middle aged to elderly uncles, all of whom fished muskelunge from flat-boats on the Allegheny River-- all of whom fished the trout streams that flowed on freestone mountains. He learned gunmanship and hunting from veterans of the Great War-- learned fishing tactics from those who knew intimately the geology of the oil lands.

Aunts Rosamond and Coletta had overseen much of his early upbringing, filling in for both father and working mother in times when that was needed. My father was an altar boy in the Eldred Church. He was schooled in Latin and in Civics. Rosamond always harbored hopes that my father might go to Saint Bonaventure University. It seems that she had many hopes that my Dad might go into teaching, and that he somehow might help reunite our side of the Cawleys with those remaining of the Catholic priest Cauleys of Erie and Buffalo.

Instead, coming of age, he enrolled at the Mansfield State Teaching College, and began a major in Science and Physics. He was interested in the world, and far more interested in the natural and physical sciences than he was in history and rituals of the Mother Church.

He had friends, colleagues and companions. He liked the technical science laboratories, and access to the college library. But I am also led to believe that he was homesick, and missed his life and status within the small town of Eldred.

The family version of the story is that one of Terry's best friends, who was also his room-mate over a

Jon C. Cawley

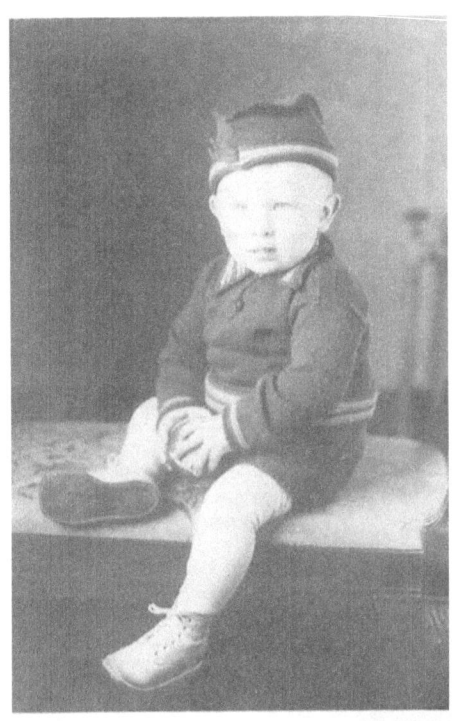

Terence J. Cawley
Baby photos.

couple of seasons, began to experience stomach ache and indigestion as senior year began. An eventual trip to the doctor revealed advanced stage stomach cancer-- and the young fellow had died within a few weeks.

    My father, whose own father had died when he was nine years old, had now seen too much of death. This event chilled hom to the core, and he left Mansfield without his Physics or Teaching degree. After a few months back in Eldred, Terence enlisted for the Air-Force instead.

    My father was smart, and good technically with his hands, and with mechanical and electrical things. He did his basic training in Michigan, including a winter season of Arctic training. Soon he was shunted into electronics, where he spent his enlisted tour wiring classified Air Force and Navy missiles, and helping to design electrical and fluidic guidance systems.

    My dad had a strange white quilted scar around one wrist that always attracted my attention which i was a child. He had reached one day into the navigational workings of a missile being tested. His metal watch bad had arced across a wire and the missile frame, welding itself in place, and heating up to very hot indeed. By the time that they got the current flow shut off, my dad had been branded by the hot watch band. He would never wear a metal watch band ever again after that.

    He was good with an oscilloscope, and good with a soldering iron. Hw developed a fascination for electro-magnetic fields, for metal detectors and mine sweepers. Upon his military stint, he had engineering certifications, and a way to make a living in the world. What he did, however, was to come home to Eldred.

    He took an engineering design job with Aerovox corporation, who was a maker of capacitors and electrical components in the region. He worked with them for a couple of years, and then he found a more suitable job with the Corning Glass Works plant in nearby Bradford.

The Bradford Oil Basin

Meanwhile, my Dad courted one Caroline J. Crain, from nearby Port Allegany. She was one year his elder, and due to a mix-up of local history, they had both graduated from the Eldred high school.

My parents married, and soon bought a small Sears Roebuck home up on a hillside at Indian Creek, about three miles outside of Eldred. My dad drove about 30 miles commute to work in Bradford, but he opted to keep his household still as near to Eldred as possible.

My mom had a daughter Lois Catherine from an early marriage, who had been born on August 16, 1956. I was born on September 14, 1963. And my little sister, Mary Margaret was born August 11, 1965.

The Indian Creek house was located at the end of a 3/4 mile long unpaved driveway that crossed a small rickety bridge over the creek, and continued up the hillside. This was located on the old Phalen oil lease. Even in the 1960s, the booming sound of the centralized power engines would echo up and down the valley.

We had a 1949 Willey's Jeep, and eventually an 1967 International Scout. We had two pet deer, named Sam and Lester, and we had dogs, including Annie and Euripides. Two pet woodchucks. And a cat named Trouble.

We had distant neighbors in the valley, but the nearest street lamp was a few miles away; the nearest MacDonalds and movie theater was about 30 miles away.

In the 1960s, the American Space Program was well under way, and my dad worked on engineering projects associated with Mercury, Gemini and Apollo, including design of the quartz glass windows of the capsules. He also worked on the pyroceram materials which would eventually become Corelle. He was part of the research project that produced the pyroceram cups in which the handles do not touch at the base of the cup.

Jon C. Cawley

The Fisher King

Jon C. Cawley

In July of 1969 my dad brought home a color television from Corning so that we could watch the Apollo 11 moon landing. I would point out that all of the footage from the moon was in black and white! We watched most of the three days transit from Earth to Moon with my Grandmother Cawley, and my two living great grandmothers, Louise Long and Eva Crain from my mom's side of the family.

Both of my great grandmothers had been born in 1883, and had lived through the era spanning from the flight of the Wright Brothers flier, to the Landing on the Moon.

Corning was beginning to get into glass and ceramic based electronic components, and my dad helped to design and operate the huge vertical glass drawing machines, which drew glass cane destined to become resistors, transister and capicitor components. My dad was a trouble-shooting engineer, and so he got to travel to Corning plants up and doen the Eastern Seaboard. Often we would travel with him.

He hunted for gold in Georgia, climbed Cadillac Mountain in Maine. We visited Disney World in Florida the year that it opened. We saw Lincoln's boyhood home, the Serpent Mound in Ohio, the Lighthouses of the Outer Banks of North Carolina. And then we would always return to Eldred and Indian Creek, and our Oil Basin.

My dad, with the help of my big sister Cathy, dynamited the spring behind our house one summer day, flinging up massive gouts of mud and stone and clay. He helped my sister build a human heart model out of baby food jars and fluidic devices for an annual science fair

My dad built his own radio telexcope out of wire in our back yard. and spent some months watching the hydrogen spectral line on his tube oscilloscope.

The Bradford Oil Basin

In later years, my dad continued working for Corning Glass, at the Bradford Electronics division, until it was eventually sold to Vishay, when he retired. He continued to invent, and to patent his several inventions.

He continued to live at Indian Creek, to maintain the long dirt driveway in the winter-time. He purchased a small power-fishing boat, which he took to Lake Erie and Chatauqua lake a few times. With age, he became much less of a hunter and fisherman, and much more of an ecologist, eventually trading his rifle and his fishing rod for a camera, while he still hiked the hills and valleys of the Oil-Basin.

Terence Joseph Cawley died of complications of cancer on June 5th, 2004.

My father's self-written epitath reads:

> The days of life are numbered
> The time goes swiftly by,
> The years in this grand old world
> So ever quickly fly.
>
> The friends that I have made here
> I have to leave behind
> But the memories I'll cherish
> And ever keep in mind.
>
> A new life is ahead of me
> And I must choose my own—
> Now, as an individual,
> I must travel on alone.

Bibliography:

American Petroleum Institute (1939). Finding and Producing Oil. American Petroleum Institute, Dallas, TX. FIRST EDITION. 338pp.

American Petroleum Institute (1951). Proceedings: Thirty-first Annual Meeting: American Petroleum Institute, Section IV: Production (Bulletin 237), Chicago, IL.

Andros, Stephen O. (1919). The Petroleum Handbook, The Shaw Publishing Company, Chicago, IL. 206pp.

Arnold, Rev. Frederick H. (1886). Cawley, The Regicide. Published by The Sussex Archaeological Collections Relating to the History and Antiquities of the County.

Ashburner, Charles A. (1880). The Geology of McKean County, and Its Connection with That of Cameron, Elk and Forest, The - Second Geological Survey of Pennsylvania: Report of Progress. Board for 2nd Geological Survey, Harrisburg, PA (1880). 371 pp.

Ashley, George H. (1924). Oil and Gas Development in the Northern Appalachian Fields in 1923, in Production of Petroleum in 1923. Papers Presented at the Symposium on Petroleum and Gas at the New York Meeting, February 1924. Published by American Institute of Mining and Metallurgical Engeneers, New York (1924)

Barber, Thomas and Woods, James (1971). Bradford Bordell and Kinzua, 2nd. printing 1971, The Elma Press, Elma, NY.

Barton, J. K. and Prentice, R. G. (1948). "Cost Study of a Water Flood in the Bradford Field," in The Producer's Monthly, February, 1948, Bradford, PA.

Bell of Pennsylvania (1980). Bradford and Surrounding Areas (telephone directory), Bell of Pennsylvania, PA.

Blaski, Inez (1980). "Railroad brought new vitality to isolated Eldred," second in a series, published in The Bradford Era, May 27, 1980, Bradford, PA.

Bovaird & Company (c. 1925). various advertising fliers, Bradford, PA.

Bovaird & Company (1936). Bovaird & Company 1936 Catalogue, Bradford, PA.

Bradford Era (1878-1881). on microfilm, Carnegie Public Library, Bradford, PA.

Brigham, Albert Perry (1911). Commercial Geography, Ginn and Company, Boston, MA. 469 pp.

Cawley, Clayton J. (c. 1925-1935) Geologic Notebooks of Clayton Cawley/Cawley Brothers Inc. Unpublished at the McKean County Historical Society. Smethport, PA.

Cawley, Jon C. (2017) The Book of Days. The Ravens Table Press/ CreateSpace Independent Publishing Platform (October 18, 2017) 474 pp.

Cawley, Jon C. (1985). Silurian Geology of Pennsylvania: A Personal interpretation. unpublished at The Pennsylvania State University, PA. 78 pp.

Cawley, William (of the Inner Temple) (1680) The Laws of Q. Elizabeth, K. James and K. C(harles the First. Concerning Jesuites, Seminary Priests, Recusants, &c. And concerning the Oaths of Supremacy and Allegiance explained by divers Judgments and Resolutions of the Reverend Judges. Together with other Observations upon the same Laws. To which is added the Statute XXV Car. II. cap. 2. for preventing dangers which may happen from Popish Recusants. And an Alphabetical Table to the whole. By William Cawley of the Inner Temple, Esq. Published by London: Printed for John Wright and Richard Chiswell at the Crown on Ludgate-Hill and the Rose and Crown in St. Paul's Church-Yard (1680). Folio.

Chilcote, Robert Harry (1953). A Geographical Analysis of Population Changes LicKean County, Pennsylvania. Masters thesis at The Pennsylvania State University, PA.

Clark, James A. (1963). The Chronological History of the Petroleum and Natural Gas Industries. Clark Book Company, Houston, TX. 317 pp.

Connelly, W. L. (1954). The Oil Business as I Saw It, Half a Century with Sinclair. University of Oklahoma Press, Norman, OK. 177 pp.

Crain, Caroline J. (2008). A Place at the Table. Caroline J. Crain Publishing, Port Allegany/Ulysses, PA. 464 pp.

Curtin, Harry H. (1951). "Treatment of Produced Water in the Bradford Field," in The Producer's Monthly, April, 1951, Bradford, PA.

Dalrymple, Sharon Ann (1994). The Cauleys Amswer God's Call. Masters Thesis.

De Golyer, E. (1940). Elements of the Petroleum Industry. The American Institute of Mining and Metallurgical Engineers, New York, NY.

Egloff, Gustav (1933), Earth Oil. A Century of Progress Series. The Williams & Wilkins Company, Baltimore, MD. 158 pp.

Environmental Protection Agency (1985). Responsiveness Summary. "Benson Reynolds Company Gas Recovery Project" (permit number 32R940BPOT).

Fettke, Charles R. (1938), The Bradford Oil Field: Pennsylvania and New York, 2nd printing 1973, Commonwealth of Pennsylvania, Harrisburg, PA.

Files of the Eldred Eagle Newspaper, at The McKean County Historical Society, Smethport PA.

Floegel, Mark (1985), "Environmental Problems Concern Oil Producers," in Olean Times Herald, October 16, 1985, Olean, NY.

Forest Oil Corporation (no date), A Short History of the Forest Oil Corporation. In-house publication, Bradford, PA.

Godcharles, Frederic A. (1933). Pennsylvania: Political, Governmental, Military Civil (in 4 volumes), The American Historical Society, Inc., New York.

Hardy, Marty Robaker (1985). "ANF Cleanup: Federal Agencies Tackling Nation's Oldest Oilfield," three part feature in The Bradford Era, September 23-25, 1985, Bradford, PA.

Harper, John A. (1981). Oil and Gas Developments in Pennsylvania in 1980 a Ten Year Review and Forecast (progress report 194), Pennsylvania Geoogical Survey, Harrisburg, PA.

Hendrick, Welland (1890). A Brief History of the Empire State. 4th ed. 1894, C.W. Bardeen, Syracuse, NY. 220 pp.

Henretta, J. E. (1929). Kane and the Upper Allegheny. Privately printed, Philadelphia, PA. 357 pp.

Herrick, John P. (1949), Empire Oil: The Story of Oil in New York State, , Mead & Company, New York, NY. 474 pp.

Hess, Terry L. (1983), "McKean County, Where Gold Is Green" in Pennsylvania Heritage, Winter, 1983, vol.9, no.2. Harrisburg, PA.

Hoskins, Donald M. (1964). Fossil Collecting in Pennsylvania (general geology report 40), 4th printing 1976, Pennsylvania Geological Survey, Harrisburg, PA.

Hungerford, Edward (1946). Men of Erie: A Story of Human Effort, Random House, New York. 346 pp.

Ireland, J. C. (1945), "White Algae," in The Producer's Monthly, September 1945, Bradford, PA.

Jaffe, Mark (1985), "Inland Spill: EPA Cleaning Up Oil Fields in PA" in The Philadelphia Inquirer, November 12, 1985, Philadelphia, PA.

Johnson, Meredith Esrey (1924). Oil and Gas Developments in Pennsylvania New York in 1924, in Production of Petroleum in 1924, AIME Symposium, New York, NY.

Kelley, Dana R. et al. (1970), The Petroleum Industry and The Future Petroleum Province in Pennsylvania in 1970 (mineral resource report 65), third printing 1983, Pennsylvania Geological Survey, Harrisburg, PA

Kennedy, Jos. C. G. (1862), Preliminary Report on the Eighth Census. 1860, Government Printing Office, Washington, DC.

Kilmer, Lawrence W. (1974), Bradford & Foster Brook Peg Leg Railroad, The Press, Elma, NY.

Lawrence, Albert A. (1938) Petroleum Comes of Age, Scott Rice Company, Tulsa, OK.

Lytle, Villiam S. (1960), History, Present Status and Future Possibilities of secondary Recovery Operations in Pennsylvania (mineral resource report 41), 2nd printing 1974, Pennsylvania Geological Survey, Harrisburg, PA.

Lytle, Villiam S. and Goth, Joseph H. (1970) Oil and Gas Geology of the Kinzua Quadrangle, Warren and McKean Counties, Pennsylvania (mineral resource report 62), 3rd printing 1982, Pennsylvania Geological Survey, Harrisburg, PA.

Martin, J. C. (1940), "Soundness" in The Producer's Monthly, February, 1940. Bradford, PA.

Martin, J. C. (1941), "Some Interesting Facts about the Bradford Field "in 'roducer's Monthly, March, 1941, Bradford, PA.

McKean County Planning Commission (1977), McKean County Economic Developement Plan (project no. ARC-DCA-76-2), County Courthouse, Smethport, PA.

McKinney, C. M. and Garton, E. L. (1957), Analyses of Crude Oils from 470 Important Oilfields in the United States (Bureau of Mines report 5376), US Government Printing Office, Washington, DC.

Morris, Edmund (1865). Derrick and Drill, Or, An Insight Into the Discovery, Development, and Present Condition and Future Prospects of Petroleum In New York, Pennsylvania, Ohio, West Virginia, &c. James Miller, New York, NY. 289 pp.

Murphy, Blakely M. (1949). Conservation of Oil & Gas: A Legal History 1948. Section of Mineral Law, American Bar Association, Chicago, IL.

O'Day, John Christopher (1906). Oil Wells in the Woods, The Oquaga Press, it, NY.

Panyity, L. S. (1920). Prospecting for Oil and Gas. John Wiley & Sons, Inc., New York, NY.

Peacey, J. T. (2004). William Cawley, Oxford DNB, 2004.

Pennsylvania State College (1930). Proceedings of the First Petroleum and Natural Gas Conference (bulletin 9), School of Mineral Industries, The Pennsylvania State College, PA.

Pennsylvania State College (1931). Proceedings of the Second Petroleum and Natural-Gas Conference Bulletin 11), School of Mineral Industries, The Pennsylvania State College, PA.

Schanz, John J. Jr. (1957). Historical Statistics of Pennsylvania's Mineral Industries, 1759-1955 (bulletin 69), School of Mineral Industries, The Pennslvania State College, PA.

Solberg, Carl (1976). Oil Power: The Rise and Imminent Fall of an American Empire, New American Library, New York, NY. 308 pp

Stahl, Drew (1981). "Petroleum and Natural Gas Engineering at PennState- Then and Now," in Earth and Mineral Sciences Bulletin, vol.50, no.3, Pennsylvania State University, State College, PA.

Stone, Rufus Barrett (1926). McKean: The Governor's County, Lewis Historical Publishing Company, New York, NY. 315 pp.

Stout, Robert E. (1972), The Kendall & Eldred Narrow Gauge Railroad, E. Stout, Allegany, NY.

Talbot, Frederick A. (1914), The Oil Conquest of the World, The J. B. Lippincot Company, Philadelphia, PA. 310 pp.

Thompson, A. Beeby (1925), Oil Field Exploration and Development, vol.2 Oil Field Practice, 2 ed. 1950. Technical Press, London.

Thornton, George W. (c. 1963), "Kinzua Viaduct Still Great Attraction," by Seneca Highlands Association and the Mt. Jewett Chamber of rce. Original by the Erie Lackawanna Railroad, NY.

Turnbull, George F. (1925), Review of Appalachian Fields for 1925 in Production of Petroleum in 1925, AIME Symposium, New York, NY.

Turnbull, George F. (1926), Review of Appalachian Fields for 1926 in Production of Petroleum in 1926, AIME Symposium, New York, NY.

U.S. Steel Company (1937). U.S. Steel News, Oil Well Supply Company Number, May, 1937, vol.2, no.5, special issue on oil. U. S. Steel Company, Hoboken, NJ.

Watts, Ralph L. (1925). Rural Pennsylvania, The Macmillan Company, New York, NY. 331 pp.

Winchester, James H. (1981). A Century of Leadership: A History of Kendall Refining Company 1881-1981, Kendall Refining Company, Bradford, PA.

Whitman, Benjamin (1896). Nelson's biographical dictionary and historical reference book of Erie County, Pennsylvania : containing a condensed history of Pennsylvania, of Erie County, and of the several cities, boroughs and townships in the county; also portraits and biographies of the governors since 1790, and of numerous representative citizens / historical and descriptive matter prepared by Benjamin Whitman., v.1.

Writers Program of Pennsylvania (1957). Pennsylvania: A Guidebook to the Keystone State. The Pennsylvania Historical Commission and others; Oxford University Press.

Wyer, Samual S. (1929). The Smithsonian Institution's Study of Natural Resources Applied to Pennsylvania's Resources. Columbus, OH.

(1940) "Pumping in Bradford Field", in The Producer's Monthly, July, 1940, Bradford, PA.

(1945) "Penn-Grade Secondary Recovery Laboratory," in The Producer's Monthly, January, 1945, Bradford, PA.

(1953). "Forgotten Oil Well Drilling Methods" in Tomorrow's Tools Today, vol. 0.4, Lane Wells Company, Los Angeles, CA.

(1960). 1960 Pennsylvania Statistical Abstract, Department of Internal Affairs, Bureau of Statistics, Harrisburg, PA.

(1971). 1971 Pennsylvania Statistical Abstract, Department of Internal rs, Bureau of Statistics, Harrisburg, PA.

(c. 1977). Index of The McKean County Miner, 1873-1890, at Bradford Library. Museum staff, Smethport, PA.

(1980). 1980 Pennsylvania Abstract, Department of Commerce, Commonwealth of Pennsylvania, Harrisburg, PA.

(1980). Pennsylvania Industrial Census, Pennsylvania Department of Commerce, Harrisburg, PA.

www.ingramcontent.com/pod-product-compliance
Lightning Source LLC
Chambersburg PA
CBHW020626220526
45464CB00001B/35